HIGHER
GCSE MATHEMATICS FOR OCR

Howard Baxter, Michael Handbury,
John Jeskins, Jean Matthews,
Mark Patmore

HOMEWORK BOOK

Hodder Murray
www.hoddereducation.co.uk

Series Editor
Brian Seager

Hodder Headline's policy is to use papers that are natural, renewable and recyclable products and made from wood grown in sustainable forests. The logging and manufacturing processes are expected to conform to the environmental regulations of the country of origin.

Orders: please contact Bookpoint Ltd, 130 Milton Park, Abingdon, Oxon OX14 4SB. Telephone: (44) 01235 827720. Fax: (44) 01235 400454. Lines are open from 9 a.m. to 5 p.m., Monday to Saturday, with a 24-hour message-answering service. Visit our website at www.hoddereducation.co.uk.

First published in 2006 by
Hodder Murray, an imprint of Hodder Education,
a member of the Hodder Headline Group,
338 Euston Road,
London, NW1 3BH

Impression number   10 9 8 7 6 5 4 3 2 1
Year   2011 2010 2009 2008 2007 2006

Cover photo © Charles D. Winters/Science Photo Library
Typeset in 10/12 Times by Tech-Set Ltd, Gateshead, Tyne & Wear
Printed by Martins The Printers, Berwick-upon-Tweed.

A catalogue record for this title is available from the British Library

ISBN-10: 0340 889 519
ISBN-13: 978 0340 889 510

→ CONTENTS

→ INTRODUCTION

This book contains exercises designed to be used for the Higher tier of GCSE Mathematics. It is particularly aimed at the OCR Specification A and each exercise matches one in the OCR Higher Student's Book.

In the Homework Book, the corresponding exercises have the same number and end in H. Thus, for example, if you have been working on Further percentages in class and used Exercise 22.1, then the homework exercise is 22.1H. The homework exercises cover the same mathematics.

Some questions are intended to be completed without a calculator, just as in the Student's Book. These are shown with a non-calculator icon in the same way. Doing these questions without a calculator is vital preparation for the non-calculator section of the examination papers.

These homework exercises provide extra practice and are also in a smaller book to carry home! If you have understood the topics, you should be able to tackle these exercises confidently as they are no harder than those you have done in class and in some cases may be a little easier. See if you agree.

More practice helps to reinforce the ideas you have learned and makes it easier to remember at a later stage. If, however, you do forget, further help is at hand. As well as the Student's Book, there is also, with this book, a Personal Tutor CD-ROM. This contains worked examples on key topics to help you recall what you have forgotten. Look out for this symbol pt next to an exercise, which shows you that there is a relevant example on the CD-ROM.

Brian Seager
Series Editor

1 → INTEGERS, POWERS AND ROOTS

EXERCISE 1.1H

Write each of these numbers as a product of its prime factors.

1	14	**2**	16
3	28	**4**	35
5	42	**6**	49
7	108	**8**	156
9	225	**10**	424

EXERCISE 1.2H

For each of these pairs of numbers
- write the numbers as products of their prime factors.
- state the highest common factor.
- state the lowest common multiple.

1	6 and 8	**2**	8 and 18
3	15 and 25	**4**	36 and 48
5	25 and 55	**6**	33 and 55
7	54 and 72	**8**	30 and 40
9	45 and 63	**10**	24 and 50

EXERCISE 1.3H

Work out these.

1	2×3	**2**	-5×8
3	-6×-2	**4**	-4×6
5	5×-7	**6**	-3×7
7	-4×-5	**8**	$28 \div -7$
9	$-25 \div 5$	**10**	$-20 \div 4$
11	$24 \div 6$	**12**	$-15 \div -3$
13	$-35 \div 7$	**14**	$64 \div -8$
15	$27 \div -9$		
16	$3 \times 6 \div -9$		
17	$-42 \div -7 \times -3$		
18	$5 \times 6 \div -10$		
19	$-9 \times 4 \div -6$		
20	$-5 \times 6 \times -4 \div -8$		

EXERCISE 1.4H

Do not use your calculator for questions **1** and **2**.

1 Write down the value of each of these.

    (a) 1^2 **(b)** 13^2

    (c) $\sqrt{64}$ **(d)** $\sqrt{196}$

    (e) 3^3 **(f)** 5^3

    (g) $\sqrt[3]{8}$ **(h)** $\sqrt[3]{64}$

2 A cube has sides of length 4 cm.
What is its volume?

You may use your calculator for questions **3** to **7**.

3 Find the square of each of these numbers.

    (a) 20   **(b)** 42   **(c)** 5.1

    (d) 60   **(e)** 0.9

4 Find the cube of each of these numbers.

    (a) 7   **(b)** 3.5   **(c)** 9.4

    (d) 20   **(e)** 100

5 Find the square root of each of these numbers. Where necessary, give your answer correct to 2 decimal places.

    (a) 900   **(b)** 75   **(c)** 284

    (d) 31 684   **(e)** 40 401

6 Find the cube root of each of these numbers. Where necessary, give your answer correct to 2 decimal places.

    (a) 729   **(b)** 144   **(c)** 9.261

    (d) 4848   **(e)** 100 000

7 A square has an area of 80 cm².
What is the length of one side? Give your answer correct to 2 decimal places.

EXERCISE 1.5H

1 Write these in simpler form using indices.
   (a) $2 \times 2 \times 2 \times 2 \times 2 \times 2$
   (b) $7 \times 7 \times 7 \times 7$
   (c) $2 \times 2 \times 3 \times 3 \times 3 \times 3 \times 5 \times 5 \times 5$
       Hint: Write the different numbers separately.

2 Work out these giving your answers in index form.
   (a) $2^2 \times 2^4$ (b) $3^6 \times 3^2$
   (c) $4^2 \times 4^3$ (d) $5^6 \times 5$

3 Work out these giving the answers in index form.
   (a) $5^5 \div 5^2$ (b) $7^8 \div 7^2$
   (c) $2^6 \div 2^4$ (d) $3^7 \div 3^3$

4 Work out these giving your answers in index form.
   (a) $5^5 \times 5^3 \div 5^2$ (b) $10^4 \times 10^6 \div 10^5$
   (c) $8^3 \times 8^3 \div 8^4$ (d) $3^5 \times 3 \div 3^3$

5 Work out these giving your answers in index form.
   (a) $\dfrac{2^5 \times 2^4}{2^3}$ (b) $\dfrac{3^7}{3^5 \times 3^2}$
   (c) $\dfrac{5^5 \times 5^4}{5^2 \times 5^3}$ (d) $\dfrac{7^5 \times 7^2}{7^2 \times 7^4}$

EXERCISE 1.6H

Do not use your calculator for questions **1** to **3**.

1 Write down the reciprocal of each of these numbers.
   (a) 4   (b) 9   (c) 65
   (d) 10   (e) 4.5

2 Write down the numbers of which these are the reciprocals.
   (a) $\frac{1}{6}$ (b) $\frac{1}{10}$ (c) $\frac{1}{25}$
   (d) $\frac{1}{71}$ (e) $\frac{2}{15}$

3 Find the reciprocal of each of these numbers. Give your answers as fractions or mixed numbers.
   (a) $\frac{3}{5}$ (b) $\frac{4}{9}$ (c) $2\frac{2}{5}$
   (d) $5\frac{1}{3}$ (e) $\frac{3}{100}$

You may use your calculator for question **4**.

4 Find the reciprocal of each of these numbers. Give your answers as decimals.
   (a) 25   (b) 0.2   (c) 6.4
   (d) 625   (e) 0.16

2 → ALGEBRAIC MANIPULATION 1

EXERCISE 2.1H

Expand these.

1 $7(3a + 6b)$
2 $5(2c + 3d)$
3 $4(3e - 5f)$
4 $3(7g - 2h)$
5 $3(4i + 2j - 3k)$
6 $3(5m - 2n + 3p)$
7 $6(4r - 3s - 2t)$
8 $8(4r + 2s + t)$
9 $4(3u + 5v)$
10 $6(4w + 3x)$
11 $2(5y + z)$
12 $4(3y + 2z)$
13 $5(3v + 2)$
14 $3(7 + 4w)$
15 $5(1 - 3a)$
16 $3(8g - 5)$

EXERCISE 2.2H

Expand the brackets and simplify these.

1 **(a)** $3(4a + 5) + 2(3a + 4)$
   (b) $5(4b + 3) + 3(2b + 1)$
   (c) $2(3 + 6c) + 4(5 + 7c)$
2 **(a)** $2(4x + 5) + 3(5x - 2)$
   (b) $4(3y + 2) + 5(3y - 2)$
   (c) $3(4 + 7z) + 2(3 - 5z)$
3 **(a)** $4(4s + 3t) + 5(2s + 3t)$
   (b) $3(4v + 5w) + 2(3v + 2w)$
   (c) $6(2x + 5y) + 3(4x + 2y)$
   (d) $2(5v + 4w) + 3(2v + w)$
4 **(a)** $5(2n + 5p) + 4(2n - 5p)$
   (b) $3(4q + 6r) + 5(2q - 3r)$
   (c) $7(3d + 2e) + 5(3d - 2e)$
   (d) $5(3f + 8g) + 4(3f - 9g)$
   (e) $4(5h - 6j) - 6(2h - 5j)$
   (f) $4(5k - 6m) - 3(2k - 5m)$

EXERCISE 2.3H

Factorise these.

1 **(a)** $8x + 20$
   (b) $3x + 6$
   (c) $9x - 12$
   (d) $5x - 30$
2 **(a)** $16 + 8x$
   (b) $9 + 15x$
   (c) $12 - 16x$
   (d) $8 - 12x$

3 **(a)** $4x^2 + 16x$
   (b) $6x^2 + 30x$
   (c) $8x^2 - 20x$
   (d) $9x^2 - 15x$

EXERCISE 2.4H

Expand the brackets and simplify these.

1 **(a)** $(a + 5)(a + 3)$
   (b) $(b + 2)(b + 4)$
   (c) $(4 + c)(3 + c)$
2 **(a)** $(2d + 3)(4d - 3)$
   (b) $(5e + 4)(3e - 2)$
   (c) $(3 + 8f)(2 - 5f)$
3 **(a)** $(3g - 2)(5g - 6)$
   (b) $(4h - 5)(3h - 7)$
   (c) $(5j - 6)(3j - 8)$
4 **(a)** $(3k + 7)(4k - 5)$
   (b) $(2 + 7m)(3 - 8m)$
   (c) $(4 + 3n)(2 - 5n)$
5 **(a)** $(2 + 5p)(3 - 7p)$
   (b) $(5r - 6)(2r - 5)$
   (c) $(3s - 2)(4s - 9)$

EXERCISE 2.5H

Simplify each of the following, writing your answer using index notation.

1 **(a)** $7 \times 7 \times 7 \times 7 \times 7$
   (b) $3 \times 3 \times 3 \times 3 \times 3$
   (c) $2 \times 2 \times 2 \times 2 \times 2 \times 2$
2 **(a)** $d \times d \times d \times d \times d \times d \times d$
   (b) $m \times m \times m \times m \times m \times m$
   (c) $t \times t \times t \times t \times t \times t \times t$
3 **(a)** $a \times a \times a \times a \times b \times b$
   (b) $c \times c \times c \times c \times d \times d \times d \times d \times d$
   (c) $r \times r \times r \times s \times s \times t \times t \times t \times t$
4 **(a)** $2x \times 3y \times 6z$
   (b) $2a \times 3b \times 4c$
   (c) $r \times 2s \times 3t \times 4s \times 5r$

EXERCISE 3.1H

Find the area of each of these triangles.

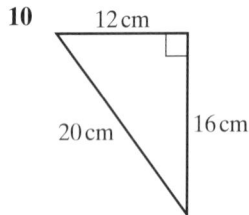

1

5 cm
6 cm

2
10 cm
8 cm

3
4 cm
9 cm

4
8 m
5 m

5
15 cm
8 cm

6
20 cm
16 cm

7
12 mm
16 mm

8
3 cm
7 cm

9
8 cm
7 cm
11 cm

10
12 cm
20 cm
16 cm

EXERCISE 3.2H

Find the area of each of these parallelograms.

1
4 cm
7 cm

2
6 cm
9 cm

3
10 cm
2.5 cm

4
7 m
8 m

5
5 cm
6 cm
4 cm

6
4 cm
5.5 cm

7
10 cm
5 cm
11 cm

8
7 cm
6 cm
12 cm

Find the missing length in each of these diagrams.

9

h cm

5 cm

Area = 40 cm^2

10

x cm

4 cm

Area = 36 cm^2

EXERCISE 3.3H

Find the size of the lettered angles. Give a reason for each answer.

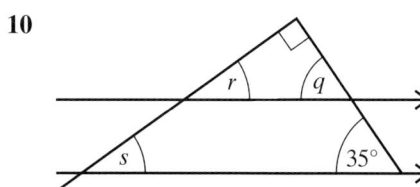

1

a

60°

2

105°

b

3

65°

c

4

49°

d

e

5

f g

120°

6

100°

h i

7

k

j

60°

8

n

l

m

118°

9

40° 133°

p o

10

r q

s 35°

EXERCISE 3.4H

Find the size of the lettered angles. Give a reason for each answer.

1

2

3

4

5

6

7

8

9

10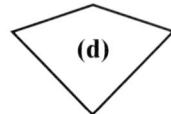

EXERCISE 3.5H

1   Name each of these quadrilaterals.

2 Name the quadrilateral or quadrilaterals which have the following properties.
  (a) Four right angles
  (b) Both pairs of opposite sides parallel
  (c) Equal diagonals
  (d) At least one pair of opposite sides parallel
  (e) Diagonals that bisect each other

3 Plot each set of points on squared paper and join them in order to make quadrilateral. Use a different grid for each part.
  Write down the special name of each quadrilateral.
  (a) (3, 0), (5, 2), (3, 4), (1, 2)
  (b) (2, 1), (4, 1), (4, 5), (2, 5)
  (c) (1, 2), (3, 1), (3, 6), (1, 7)
  (d) (2, 1), (2, 5), (8, 2), (8, 4)

4 A rhombus is a special type of parallelogram.
  What extra properties does a rhombus have?

5 A quadrilateral has angles of 80°, 100°, 80°, 100° in order as you work your way around the quadrilateral. The sides are not all the same length.
  Which special quadrilateral could have these as their angles?
  Draw the quadrilateral and mark on the angles.

EXERCISE 3.6H

1 A polygon has nine sides.
  Work out the sum of the interior angles of this polygon.

2 A polygon has 13 sides.
  Work out the sum of the interior angles of this polygon.

3 Four of the exterior angles of a hexagon are 93°, 50°, 37° and 85°.
  The other two angles are equal.
  (a) Work out the size of these equal exterior angles.
  (b) Work out the interior angles of the hexagon.

4 Four of the interior angles of a pentagon are 170°, 80°, 157°, and 75°.
  (a) Work out the size of the other interior angle.
  (b) Work out the exterior angles of the pentagon.

5 A regular polygon has 18 sides.
  Find the size of the exterior and the interior angle of this polygon.

6 A regular polygon has 24 sides.
  Find the size of the exterior and the interior angle of this polygon.

7 A regular polygon has an exterior angle of 12°.
  Work out the number of sides that the polygon has.

8 A regular polygon has an interior angle of 172°.
  Work out the number of sides that the polygon has.

4 → FRACTIONS, DECIMALS AND PERCENTAGES

EXERCISE 4.1H

1 For each pair of fractions
- find the common denominator.
- state which is the bigger fraction.

   (a) $\frac{7}{8}$ or $\frac{3}{4}$ **(b)** $\frac{5}{9}$ or $\frac{7}{11}$ **(c)** $\frac{1}{6}$ or $\frac{3}{20}$

2 Work out these.

   (a) $\frac{3}{7} + \frac{2}{7}$ **(b)** $\frac{7}{15} + \frac{4}{15}$ **(c)** $\frac{8}{11} - \frac{3}{11}$

   (d) $\frac{11}{17} - \frac{8}{17}$ **(e)** $\frac{7}{16} + \frac{3}{16}$ **(f)** $\frac{7}{9} + \frac{4}{9}$

   (g) $\frac{7}{12} - \frac{5}{12}$ **(h)** $\frac{8}{11} + \frac{5}{11}$ **(i)** $2\frac{4}{7} + 3\frac{1}{7}$

   (j) $4\frac{5}{6} - 1\frac{1}{6}$ **(k)** $5\frac{9}{13} - \frac{4}{13}$ **(l)** $4\frac{3}{8} - 1\frac{5}{8}$

3 Work out these.

   (a) $\frac{2}{9} + \frac{1}{3}$ **(b)** $\frac{7}{12} + \frac{1}{4}$ **(c)** $\frac{3}{4} - \frac{1}{10}$

   (d) $\frac{13}{16} - \frac{3}{8}$ **(e)** $\frac{7}{8} + \frac{1}{3}$ **(f)** $\frac{4}{5} + \frac{5}{6}$

   (g) $\frac{7}{12} - \frac{1}{8}$ **(h)** $\frac{9}{20} + \frac{3}{4}$ **(i)** $\frac{7}{11} + \frac{3}{5}$

   (j) $\frac{7}{12} + \frac{7}{10}$ **(k)** $\frac{7}{8} - \frac{1}{6}$ **(l)** $\frac{7}{15} - \frac{3}{20}$

4 Work out these.

   (a) $4\frac{1}{4} + 3\frac{1}{3}$ **(b)** $6\frac{8}{9} - 1\frac{2}{3}$ **(c)** $5\frac{3}{8} + \frac{1}{4}$

   (d) $5\frac{11}{16} - 2\frac{1}{8}$ **(e)** $2\frac{5}{6} + 3\frac{1}{4}$ **(f)** $6\frac{8}{9} - 2\frac{1}{6}$

   (g) $3\frac{5}{8} + 4\frac{7}{10}$ **(h)** $5\frac{7}{11} - 5\frac{1}{3}$ **(i)** $4\frac{3}{4} + 3\frac{2}{7}$

   (j) $6\frac{1}{4} - 2\frac{2}{3}$ **(k)** $7\frac{1}{9} - 2\frac{1}{2}$ **(l)** $5\frac{3}{10} - 4\frac{4}{5}$

EXERCISE 4.2H

1 Change these mixed numbers to improper fractions.

   (a) $4\frac{3}{5}$ **(b)** $6\frac{1}{4}$ **(c)** $3\frac{4}{7}$ **(d)** $1\frac{5}{9}$

   (e) $4\frac{5}{6}$ **(f)** $7\frac{3}{10}$ **(g)** $4\frac{7}{8}$

2 Work out these.
Write your answers as proper fractions or mixed numbers in their lowest terms.

   (a) $\frac{3}{7} \times 5$ **(b)** $\frac{5}{9} \times 6$ **(c)** $\frac{3}{5} \div 4$

   (d) $6 \times \frac{5}{11}$ **(e)** $\frac{2}{9} \div 4$ **(f)** $9 \div \frac{3}{8}$

3 Work out these.
Write your answers as proper fractions or mixed numbers in their lowest terms.

   (a) $\frac{2}{3} \times \frac{5}{7}$ **(b)** $\frac{1}{8} \times \frac{5}{6}$ **(c)** $\frac{7}{9} \times \frac{2}{5}$

   (d) $\frac{5}{8} \div \frac{3}{4}$ **(e)** $\frac{3}{8} \div \frac{1}{3}$ **(f)** $\frac{4}{9} \times \frac{5}{11}$

   (g) $\frac{6}{7} \times \frac{1}{8}$ **(h)** $\frac{7}{15} \div \frac{2}{3}$ **(i)** $\frac{7}{12} \times \frac{3}{8}$

   (j) $\frac{9}{16} \div \frac{7}{12}$ **(k)** $\frac{7}{10} \div \frac{5}{12}$ **(l)** $\frac{7}{30} \times \frac{10}{21}$

4 Work out these.
Write your answers as proper fractions or mixed numbers in their lowest terms.

   (a) $4\frac{3}{4} \times 1\frac{7}{9}$ **(b)** $3\frac{2}{3} \times \frac{1}{5}$ **(c)** $4\frac{2}{5} \div 2\frac{4}{5}$

   (d) $1\frac{3}{11} \div 3\frac{1}{2}$ **(e)** $4\frac{1}{2} \times 3\frac{2}{3}$ **(f)** $3\frac{5}{9} \div 2\frac{2}{3}$

   (g) $3\frac{2}{7} \times 1\frac{5}{9}$ **(h)** $2\frac{5}{8} \div 1\frac{5}{6}$ **(i)** $1\frac{7}{15} \times 12\frac{1}{2}$

   (j) $5\frac{3}{5} \div 1\frac{3}{4}$ **(k)** $6\frac{2}{9} \times 2\frac{1}{8}$ **(l)** $7\frac{1}{2} \div 2\frac{3}{5}$

EXERCISE 4.3H

1 Work out these.

(a) $\frac{3}{4} + \frac{1}{6}$ (b) $\frac{5}{8} - \frac{2}{7}$

(c) $\frac{5}{9} \times \frac{3}{8}$ (d) $\frac{7}{16} \div \frac{5}{12}$

(e) $1\frac{4}{5} + 2\frac{3}{4}$ (f) $6\frac{3}{7} - 2\frac{1}{3}$

(g) $5\frac{3}{5} \times 4$ (h) $4\frac{5}{9} \div 1\frac{1}{6}$

2 Write these fractions in their lowest terms.

(a) $\frac{40}{125}$ (b) $\frac{28}{49}$ (c) $\frac{72}{192}$

(d) $\frac{225}{350}$ (e) $\frac{17}{153}$

3 Write these improper fractions as mixed numbers.

(a) $\frac{120}{72}$ (b) $\frac{150}{13}$ (c) $\frac{86}{19}$

(d) $\frac{192}{54}$ (e) $\frac{302}{17}$

4 Calculate
(a) the perimeter of this rectangle.
(b) the area of this rectangle.

$5\frac{3}{5}$ cm

$2\frac{1}{4}$ cm

EXERCISE 4.4H

1 Change each of these fractions to a decimal. If necessary, give your answer to 3 decimal places.

(a) $\frac{7}{8}$ (b) $\frac{7}{100}$ (c) $\frac{5}{9}$ (d) $\frac{2}{11}$

2 State whether each of these fractions gives a recurring or a terminating decimal. Give your reasons.

(a) $\frac{3}{4}$ (b) $\frac{5}{6}$ (c) $\frac{5}{11}$

(d) $\frac{2}{25}$ (e) $\frac{7}{32}$

3 (a) Find the recurring decimal equivalent to $\frac{3}{101}$.
(b) How many digits are there in the repeating pattern?

EXERCISE 4.5H

Work out these. As far as possible, write down only your final answer.

1	3.4 + 6.1	**2**	4.3 + 3.6
3	5.8 − 2.3	**4**	7.9 − 4.4
5	3.7 + 2.6	**6**	5.8 + 3.4
7	7.2 − 0.9	**8**	5.4 − 3.5
9	8.6 + 2.7	**10**	6.9 + 4.6
11	6.7 − 5.8	**12**	6.3 − 2.9

EXERCISE 4.6H pt

1 Work out these.

(a) 5×0.4 (b) 0.6×8

(c) 4×0.7 (d) 0.9×6

(e) 0.7×0.3 (f) 0.9×0.4

(g) 50×0.7 (h) 0.4×80

(i) 0.7×0.1 (j) 0.4×0.2

(k) $(0.8)^2$ (l) $(0.2)^2$

2 Work out these.

(a) $6 \div 0.3$ (b) $4.8 \div 0.2$

(c) $2.4 \div 0.6$ (d) $7.2 \div 0.4$

(e) $33 \div 1.1$ (f) $60 \div 1.5$

(g) $12 \div 0.4$ (h) $35 \div 0.7$

(i) $64 \div 0.8$ (j) $32 \div 0.2$

(k) $2.17 \div 0.7$ (l) $47.5 \div 0.5$

3 Work out these.

(a) 3.6×1.4 (b) 5.8×2.6

(c) 8.1×4.3 (d) 6.5×3.2

(e) 74×1.7 (f) 64×3.8

(g) 2.9×7.6 (h) 11.4×3.2

(i) 25.2×0.8 (j) 2.67×0.9

(k) 8.45×1.2 (l) 7.26×2.4

4 Work out these.

(a) $23.6 \div 0.4$ (b) $23.4 \div 0.8$

(c) $18.2 \div 0.7$ (d) $31.2 \div 0.6$

(e) $42.3 \div 0.9$ (f) $75.6 \div 1.2$

(g) $5.28 \div 0.3$ (h) $7.56 \div 0.7$

(i) $63.2 \div 0.2$ (j) $6.27 \div 1.1$

(k) $3.51 \div 1.3$ (l) $8.19 \div 1.3$

EXERCISE 4.7H 🅟

1   Write down the multiplier that will increase an
    amount by
    (a) 17%.   **(b)** 30%.   **(c)** 73%.
    (d) 6%.   **(e)** 1%.   **(f)** 12.5%.
    (g) 160%.

2   Write down the multiplier that will decrease
    an amount by
    (a) 13%.   **(b)** 40%.   **(c)** 35%.
    (d) 8%.   **(e)** 4%.   **(f)** 27%.
    (g) 15.5%.

3   Mrs Green bought an antique for £200.
    She later sold it at 250% profit.
    What did she sell it for?

4   Jane earns £14 500 per year.
    She receives an increase of 2%.
    Find her new salary.

5   In a sale all items are reduced by 20%.
    Shamir bought a computer in the sale.
    The original price was £490.
    What was the sale price?

6   Graham invested £3500 at 4% compound
    interest.
    What was the investment worth at the end of
    5 years?
    Give your answer to the nearest pound.

7   A car decreased in value by 11% per year.
    If it cost £16 500 new, what was it worth after
    4 years?
    Give your answer to a suitable degree of
    accuracy.

8   In a certain country the population rose by 5%
    every year from 1999 to 2004.
    If the population was 26.5 million in 1999,
    what was the population in 2004?
    Give your answer in millions to the nearest 0.1
    of a million.

9   Jane invested £4500 with compound interest
    for 3 years.
    She could receive either 3% interest every
    6 months or 6% interest every year.
    Which should Jane choose?
    How much more will she receive?

10 Prices went up by 2% in 2003, 3% in 2004
    and 2.5% in 2005.
    If an item cost £32 at the start of 2003 what
    would it cost at the end of 2005?

EXERCISE 5.1H

1   For a project, Rebecca recorded the ages of 100 cars as they passed the school gates one morning. Here are her results.

Age (a years)	$0 \leqslant a < 2$	$2 \leqslant a < 4$	$4 \leqslant a < 6$	$6 \leqslant a < 8$	$8 \leqslant a < 10$	$10 \leqslant a < 12$	$12 \leqslant a < 14$
Frequency	16	23	24	17	12	7	1

   (a) Draw a frequency diagram to show these data.
   (b) Which of the intervals is the modal group?

2   The manager of a leisure centre recorded the weights of 120 men. Here are the results.

Weight (w kg)	$60 \leqslant w < 65$	$65 \leqslant w < 70$	$70 \leqslant w < 75$	$75 \leqslant w < 80$	$80 \leqslant w < 85$	$85 \leqslant w < 90$
Frequency	4	18	36	50	10	2

   (a) Draw a frequency diagram to represent these data.
   (b) Which of the intervals is the modal group?
   (c) Which of the intervals contains the median value?

3   This frequency diagram shows the times taken by a group of girls to run a race.
   (a) How many girls took longer than 9 minutes?
   (b) How many girls took part in the race?
   (c) What percentage of the girls took less than 7 minutes?
   (d) What is the modal finishing time?
   (e) Use the diagram to draw up a grouped frequency table like those in questions 1 and 2.

EXERCISE 5.2H

1 The table shows the heights of 40 plants.

Height (h cm)	$3 \leqslant h < 4$	$4 \leqslant h < 5$	$5 \leqslant h < 6$	$6 \leqslant h < 7$	$7 \leqslant h < 8$	$8 \leqslant h < 9$
Frequency	1	7	10	12	8	2

Draw a frequency polygon to show these data.

2 The table shows the time taken for a group of children to get from home to school.

Time (t mins)	$0 \leqslant t < 5$	$5 \leqslant t < 10$	$10 \leqslant t < 15$	$15 \leqslant t < 20$	$20 \leqslant t < 25$	$25 \leqslant t < 30$
Frequency	3	15	27	34	19	2

Draw a frequency polygon to show these data.

3 The ages of all of the people under 70 in a small village were recorded in 1985 and 2005.
The results are given in the table below.

Age (a years)	$0 \leqslant a < 10$	$10 \leqslant a < 20$	$20 \leqslant a < 30$	$30 \leqslant a < 40$	$40 \leqslant a < 50$	$50 \leqslant a < 60$	$60 \leqslant a < 70$
Frequency 1985	85	78	70	53	40	28	18
Frequency 2005	50	51	78	76	62	64	56

(a) On the same grid, draw a frequency polygon for each year.
(b) Use the diagram to compare the distribution of ages in the two years.

EXERCISE 5.3H

1 As part of a survey, Emma measured the heights, in centimetres, of the 50 teachers in her school. Here are her results.

168	194	156	167	177	180	188	172	170	169
174	178	186	174	166	165	159	173	185	162
163	174	180	184	173	182	161	176	170	169
178	175	172	179	173	162	177	176	184	191
181	165	163	185	178	175	182	164	179	168

Construct a stem-and-leaf diagram to show these heights.

2 A group of students took a Maths test. Here are their marks.

51	94	56	45	70	67	69	49	55	71
52	73	64	60	71	58	64	35	31	81
41	67	64	73	82	57	75	33	88	79
66	52	48

(a) Construct a stem-and-leaf diagram to show these marks.
(b) The pass mark was 40. How many students had to do a re-test?
(c) What is the modal mark?
(d) Use your table to find the median mark.

3 The data shows the length of the throws, in metres, in a school shot putt competition.

4.4	5.4	8.5	9.2	7.3	5.7	9.9	9.6	7.4	9.1
7.5	7.0	8.3	4.9	5.7	6.4	6.7	7.3	8.2	9.0
5.2	7.0	8.9	9.1	5.2	6.4	7.3	8.2	5.9	5.0
6.5	7.9	8.2	8.9	5.3	5.2

(a) Construct a stem-and-leaf diagram to show these lengths.
(a) How many people took part in the competition?
(a) Use your table to find the median length.

EXERCISE 5.4H

1 Bill grows tomatoes. As an experiment he divided his land into eight plots.
He used a different amount of fertiliser on each plot.
The table shows the weight of tomatoes he got from each of the plots.

Amount of fertiliser (g/m²)	10	20	30	40	50	60	70	80
Weight of tomatoes (kg)	36	41	58	60	70	76	75	92

    (a) Draw a scatter diagram to show this information.
    (b) Describe the correlation shown in the scatter diagram.
    (c) The mean of the amount of fertiliser is 45 g/m². Calculate the mean weight of the tomatoes.
    (d) Plot the point that has these means as coordinates.
    (e) Draw a line of best fit on your scatter diagram.
    (f) What weight of tomatoes should Bill expect to get if he used 75 g/m²?

2 The table shows the prices and mileages of seven second-hand cars of the same model.

Price (£)	6000	3500	1000	8500	5500	3500	7000
Mileage	29 000	69 000	92 000	17 000	53 000	82 000	43 000

    (a) Draw a scatter diagram to show this information.
    (b) Describe the correlation shown in the scatter diagram.
    (c) The mean Price(£) is £5000 and the mean of the mileages is 55 000.
    (d) Plot the point that has these means as coordinates.
    (e) Draw a line of best fit on your scatter diagram.
    (f) Use your line of best fit to estimate
        (i) the price of this model of car which has covered 18 000 miles.
        (ii) the mileage of this model of car which costs £4000.

3 The heights of 10 daughters, all aged 20, and their fathers are given in the table below.

Height of father (cm)	167	168	169	171	172	172	174	175	176	182
Height of daughter (cm)	164	166	166	168	169	170	170	171	173	177

    (a) Draw a scatter diagram to show this information.
    (b) Describe the correlation shown in the scatter diagram.
    (c) The mean height of the fathers 172.6 cm. Calculate the mean height of the daughters.
    (d) Plot the point that has these means as coordinates.
    (e) Draw a line of best fit on your scatter diagram.
    (f) Use your line of best fit to estimate the height of a 20-year-old daughter whose father is 180 cm tall.

EXERCISE 6.1H

Find the circumferences of circles with these diameters.

1	8 cm	**2**	17 cm	**3**	39.2 cm
4	116 mm	**5**	5.1 m	**6**	6.32 m
7	14 cm	**8**	23 cm	**9**	78 mm
10	39 mm	**11**	4.4 m	**12**	2.75 m

EXERCISE 6.2H

1 Find the areas of circles with the following radii.

(a)	17 cm	**(b)**	23 cm
(c)	67 cm	**(d)**	43 mm
(e)	74 mm	**(f)**	32 cm
(g)	58 cm	**(h)**	4.3 cm
(i)	8.7 cm	**(j)**	47 m
(k)	1.9 m	**(l)**	2.58 m

2 Find the areas of circles with the following diameters.

(a)	18 cm	**(b)**	28 cm
(c)	68 cm	**(d)**	38 mm
(e)	78 mm	**(f)**	58 cm
(g)	46 cm	**(h)**	6.4 cm
(i)	7.6 cm	**(j)**	32 m
(k)	3.4 m	**(l)**	4.32 m

EXERCISE 6.3H

Find the area of each of these shapes.
Break them down into rectangles and right-angled triangles first.

1

2

3

4

5

6

EXERCISE 6.4H

Find the volume of each of these shapes.

1
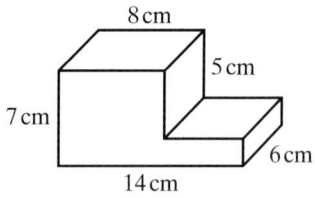
8 cm
5 cm
7 cm
6 cm
14 cm

2
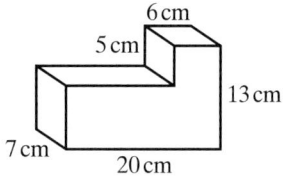
6 cm
5 cm
13 cm
7 cm
20 cm

3
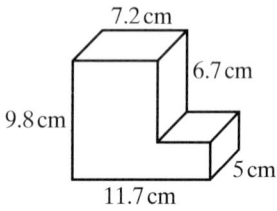
7.2 cm
6.7 cm
9.8 cm
5 cm
11.7 cm

4
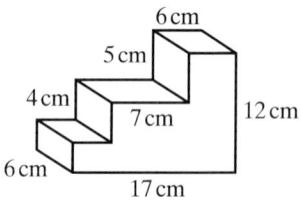
6 cm
5 cm
4 cm
7 cm
12 cm
6 cm
17 cm

5
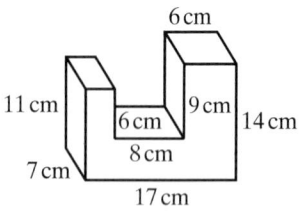
6 cm
11 cm
6 cm
9 cm
14 cm
7 cm
8 cm
17 cm

6
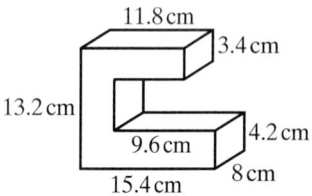
11.8 cm
3.4 cm
13.2 cm
4.2 cm
9.6 cm
8 cm
15.4 cm

EXERCISE 6.5H

Find the volume of each of these prisms.

1
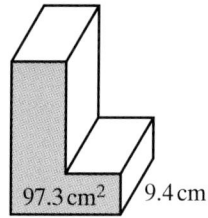
97.3 cm² 9.4 cm

2
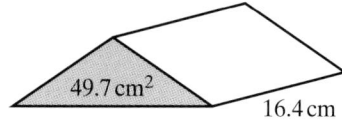
49.7 cm²
16.4 cm

3

47.1 cm
24.7 cm²

4
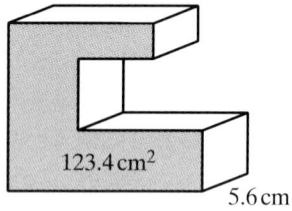
123.4 cm²
5.6 cm

5
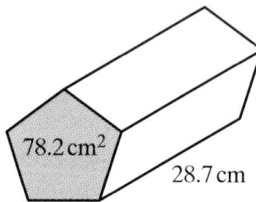
78.2 cm²
28.7 cm

6
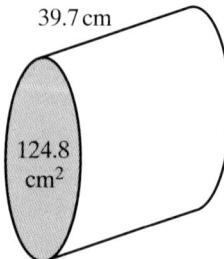
39.7 cm
124.8 cm²

EXERCISE 6.6H

Find the volumes of cylinders with these dimensions.

1 Radius 7 cm and height 29 cm
2 Radius 13 cm and height 27 cm
3 Radius 25 cm and height 80 cm
4 Radius 14 mm and height 35 mm
5 Radius 28 mm and height 8 mm
6 Radius 0.6 mm and height 5.1 mm
7 Radius 1.7 m and height 5 m
8 Radius 2.6 m and height 3.4 m

EXERCISE 6.7H

Find the curved surface areas of cylinders with these dimensions.

1 Radius 9 cm and height 16 cm
2 Radius 13 cm and height 21 cm
3 Radius 27 cm and height 12 cm
4 Radius 17 mm and height 35 mm
5 Radius 12 mm and height 6 mm
6 Radius 3.7 mm and height 63 mm
7 Radius 1.9 m and height 19 m
8 Radius 2.7 m and height 4.3 m

Find the total surface areas of cylinders with these dimensions.

9 Radius 8 cm and height 11 cm
10 Radius 17 cm and height 28 cm
11 Radius 29 cm and height 15 cm
12 Radius 32 mm and height 8 mm
13 Radius 35 mm and height 12 mm
14 Radius 3.9 mm and height 45 mm
15 Radius 0.8 m and height 7 m
16 Radius 2.9 m and height 1.7 m

EXERCISE 6.8H pt

Draw the plan view, front elevation and side elevation of each of these objects.

1

2

3

4

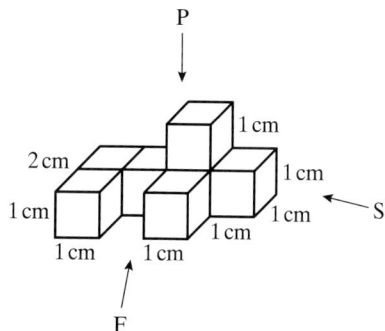

EXERCISE 7.1H

Solve these equations.

1. $2x - 3 = 7$
2. $2x + 2 = 8$
3. $2x - 9 = 3$
4. $3x - 2 = 7$
5. $6x + 2 = 26$
6. $3x + 2 = 17$
7. $4x - 5 = 3$
8. $4x + 2 = 8$
9. $2x - 7 = 10$
10. $5x + 12 = 7$
11. $x^2 + 3 = 19$
12. $x^2 - 2 = 7$
13. $y^2 - 1 = 80$
14. $11 - 3x = 2$
15. $4x - 12 = -18$

EXERCISE 7.2H

Solve these equations.

1. $3(x - 2) = 18$
2. $2(1 + x) = 8$
3. $3(x - 5) = 6$
4. $2(x + 3) = 10$
5. $5(x - 2) = 15$
6. $2(x + 3) = 10$
7. $5(x - 4) = 20$
8. $4(x + 1) = 16$
9. $2(x - 7) = 8$
10. $3(2x + 3) = 18$
11. $5(2x - 3) = 15$

12. $2(3x - 2) = 14$
13. $5(2x - 3) = 40$
14. $4(x - 3) = 6$
15. $2(2x - 3) = 8$

EXERCISE 7.3H

Solve these equations.

1. $5x - 1 = 3x + 5$
2. $5x + 1 = 2x + 13$
3. $7x - 2 = 2x + 8$
4. $6x + 1 = 4x + 21$
5. $9x - 10 = 4x + 5$
6. $5x - 8 = 3x - 6$
7. $6x + 2 = 10 - 2x$
8. $2x - 10 = 5 - 3x$
9. $15 + 3x = 2x + 18$
10. $2x - 5 = 4 - x$
11. $3x - 2 = x + 7$
12. $x - 1 = 2x - 6$
13. $2x - 4 = 2 - x$
14. $9 - x = x + 5$
15. $3x - 2 = x - 8$

EXERCISE 7.4H

Solve these equations.

1. $\dfrac{x}{2} = 7$
2. $\dfrac{x}{5} - 2 = 1$
3. $\dfrac{x}{4} + 5 = 8$
4. $\dfrac{x}{3} - 5 = 5$

5. $\dfrac{x}{6} + 3 = 4$
6. $\dfrac{x}{5} + 1 = 4$
7. $\dfrac{x}{8} - 3 = 9$
8. $\dfrac{x}{4} + 1 = 3$
9. $\dfrac{x}{7} + 5 = 6$
10. $\dfrac{x}{4} + 5 = 4$

EXERCISE 7.5H pt

For each of questions **1** to **5**, solve the inequality and show the solution on a number line.

1. $x - 2 > 1$
2. $x + 1 < 3$
3. $3x - 2 \geqslant 7$
4. $2x + 1 \leqslant 6$
5. $3x - 6 \geqslant 0$

For each of questions **6** to **15**, solve the inequality.

6. $7 \leqslant 2x - 1$
7. $5x < x + 12$
8. $4x \geqslant x + 9$
9. $4 + x < 0$
10. $3x + 1 \leqslant 2x + 6$
11. $2(x - 3) > x$
12. $5(x + 1) > 3x + 10$
13. $7x + 5 \leqslant 2x + 30$
14. $5x + 2 < 7x - 4$
15. $3(3x + 2) \geqslant 2(x + 10)$

8 ➔ RATIO AND PROPORTION

EXERCISE 8.1H

1 Write each of these ratios in its lowest terms.
   (a) 8 : 6   (b) 20 : 50   (c) 35 : 55
   (d) 8 : 24 : 32   (e) 15 : 25 : 20

2 Write each of these ratios in its lowest terms.
   (a) 200 g : 500 g
   (b) 60p : £3
   (c) 1 minute : 25 seconds
   (d) 2 m : 80 cm
   (e) 500 g : 3 kg

3 A bar of brass contains 400 g of copper and 200 g of zinc.
   Write the ratio of copper to zinc in its lowest terms.

4 Teri, Jannae and Abi receive £200, £350 and £450 respectively as their dividends in a joint investment.
   Write the ratio of their dividends in its lowest terms.

5 Three saucepans hold 500 ml, 1 litre and 2.5 litres respectively.
   Write the ratio of their capacities in its lowest terms.

EXERCISE 8.2H

1 Write each of these ratios in the form 1 : n.
   (a) 2 : 10   (b) 5 : 30
   (c) 2 : 9   (d) 4 : 9
   (e) 50 g : 30 g (f) 15p : £3
   (g) 25 cm : 6 m (h) 20 : 7
   (i) 4 mm : 1 km

2 On a map a distance of 12 mm represents a distance of 3 km.
   What is the scale of the map in the form 1 : n?

3 A picture is enlarged on a photocopier from 25 mm wide to 15 cm wide.
   What is the ratio of the picture to the enlargement in the form 1 : n?

EXERCISE 8.3H pt

1 A photo is enlarged in the ratio 1 : 5.
   (a) The length of the small photo is 15 cm.
       What is the length of the large photo?
   (b) The width of the large photo is 45 cm.
       What is the width of the small photo?

2 To make a dressing for her lawn, Rupinder mixes loam and sand in the ratio 1 : 3.
   (a) How much sand should she mix with two buckets of loam?
   (b) How much loam should she mix with 15 buckets of sand?

3 To make mortar, Fred mixes 1 part cement with 5 parts sand.
   (a) How much sand does he mix with 500 g of cement?
   (b) How much cement does he mix with 4.5 kg of sand?

4 A rectangular picture is 6 cm wide.
   It is enlarged in the ratio 1 : 4.
   How wide is the enlargement?

5 The Michelin motoring atlas of France has a scale of 1 cm to 2 km.
   (a) On the map the distance between Metz and Nancy is 25 cm.
       How far is the actual distance between the two towns?
   (b) The actual distance between Caen and Falaise is 33 km.
       How far is this on the map?

6 Graham is making pastry.
To make enough for five people he uses 300 g of flour.
How much flour should he use for eight people?

7 To make a solution of a chemical a scientist mixes 2 parts chemical with 25 parts water.
  (a) How much water should he mix with 10 ml of chemical?
  (b) How much chemical should he mix with 1 litre of water?

8 The ratio of the sides of two rectangles is 2 : 5.
  (a) The length of the small rectangle is 4 cm. How long is the big rectangle?
  (b) The width of the big rectangle is 7.5 cm. How wide is the small rectangle?

9 Jason mixes 3 parts black paint with 4 parts white paint to make dark grey paint.
  (a) How much white paint does he mix with 150 ml of black paint?
  (b) How much black paint does he mix with 1 litre of white paint?

10 In an election the number of votes was shared between the Labour, Conservatives and other parties in the ratio 5 : 4 : 2.
Labour received 7500 votes.
  (a) How many votes did the Conservatives receive?
  (b) How many votes did the other parties receive?

EXERCISE 8.4H

Do not use your calculator for questions **1** to **5**.

1 Share £40 between Paula and Tariq in the ratio 3 : 5.

2 Paint is mixed in the ratio 2 parts black paint to 3 parts white paint to make 10 litres of grey paint.
  (a) How much black paint is used?
  (b) How much white paint is used?

3 A metal alloy is made up of copper, iron and nickel in the ratio 3 : 4 : 2.
How much of each metal is there in 450 g of the alloy?

4 Inderjit worked 6 hours one day.
The time he spent on filing, writing and computing is in the ratio 2 : 3 : 7.
How long did he spend computing?

5 Daisy and Emily invested £5000 and £8000 respectively in a business venture.
They agreed to share the profits in the ratio of their investment.
Emily received £320.
What was the total profit?

You may use your calculator for questions **6** to **8**.

6 Shahida spends her pocket money on sweets, magazines and clothes in the ratio 2 : 3 : 7.
She receives £15 a week.
How much does she spend on sweets?

7 In a questionnaire the three possible answers are 'Yes', 'No' and 'Don't know'.
The answers from a group of 456 people are in the ratio 10 : 6 : 3.
How many 'Don't knows' are there?

8 Iain and Stephen bought a house between them in Spain.
Iain paid 60% of the cost and Stephen 40%.
  (a) Write the ratio of the amounts they paid in its lowest terms.
  (b) The house cost 210 000 euros. How much did each pay?

EXERCISE 8.5H

1 An 80 g bag of Munchoes costs 99p and a 200 g bag of Munchoes costs £2.19.
Show which is the better value.

2 Baxter's lemonade is sold in 2 litre bottles for £1.29 and in 3 litre bottles for £1.99.
Show which is the better value.

3 Butter is sold in 200 g tubs for 95p and in 450 g packets for £2.10. Show which is the better value.

4 Fruit yogurt is sold in packs of 4 tubs for 79p and in packs of 12 tubs for £2.19. Show which is the better value.

5 There are two packs of minced meat on the reduced price shelf at the supermarket, a 1.8 kg pack reduced to £2.50 and a 1.5 kg pack reduced to £2. Show which is the better value.

6 Smoothie shaving gel costs £1.19 for the 75 ml bottle and £2.89 for the 200 ml bottle. Show which is better value.

7 A supermarket sells cans of cola in two different sized packs: a pack of 12 cans costs £4.30 and a pack of 20 cans costs £7.25. Which pack gives the better value?

8 Sudso washing powder is sold in 3 sizes: 750 g for £3.15, 1.5 kg for £5.99 and 2.5 kg for £6.99. Which packet gives the best value?

EXERCISE 9.1H

1 For each of these sets of data
    (i) find the mode.   **(ii)** find the median.   **(iii)** find the range.   **(iv)** calculate the mean.

(a)

Score on biased dice	Number of times thrown
1	52
2	46
3	70
4	54
5	36
6	42
Total	300

(b)

Number of drawing pins in a box	Number of boxes
98	5
99	14
100	36
101	28
102	17
103	13
104	7
Total	120

(c)

Number of snacks per day	Frequency
0	23
1	68
2	39
3	21
4	10
5	3
6	1

(d)

Number of letters received on Monday	Frequency
0	19
1	37
2	18
3	24
4	12
5	5
6	2
7	3

2 Gift tokens cost £1, £5, £10, £20 or £50 each.
The frequency table below shows the numbers of each value of gift token sold in one bookstore on a Saturday.
Calculate the mean value of gift token bought in the bookstore that Saturday.

Price of gift token (£)	1	5	10	20	50
Number of tokens sold	12	34	26	9	1

3 A sample of people were asked how many visits to the cinema they had made in one month.
None of those asked had made more than eight visits to the cinema.
The table below shows the data.
Calculate the mean number of visits to the cinema.

Number of visits	0	1	2	3	4	5	6	7	8
Frequency	136	123	72	41	18	0	5	1	4

EXERCISE 9.2H

1 For each of these sets of data, calculate an estimate of
  (i) the range.
  (ii) the mean.
  (a)

Number of trains arriving late each day (x)	Number of days (f)
0–4	19
5–9	9
10–14	3
15–19	0
20–24	1
Total	32

  (b)

Number of weeds per square metre (x)	Number of square metres (f)
0–14	204
15–29	101
30–44	39
45–59	13
60–74	6
75–89	2

(c)

Number of books sold (x)	Frequency (f)
60–64	3
65–69	12
70–74	23
75–79	9
80–84	4
85–89	1

(d)

Number of days absent (x)	Frequency (f)
0–3	13
4–7	18
8–11	9
12–15	4
16–19	0
20–23	1
24–27	3

2 The table below gives the number of sentences per chapter in a book.

Number of sentences (x)	$100 \leq x < 125$	$125 \leq x < 150$	$150 \leq x < 175$	$175 \leq x < 200$	$200 \leq x < 225$
Frequency	1	9	8	5	2

(a) What is the modal group?

(b) In which group is the median number of sentences?

(c) Calculate an estimate of the mean number of sentences.

3 A group of students were asked to estimate the number of beans in a jar.
The results of their estimates are summarised in the table.
Calculate an estimate of the mean number of beans estimated by these students.

Estimated number of beans (x)	Frequency (f)
300–324	9
325–349	26
350–374	52
375–399	64
400–424	83
425–449	57
450–474	18
475–499	5

EXERCISE 9.3H

1 For each of these sets of data, calculate an estimate of
   (i)   the range.   **(ii)**   the mean.

(a)

Height of sunflower in centimetres (x)	Number of plants (f)
$100 \leqslant x < 110$	6
$110 \leqslant x < 120$	13
$120 \leqslant x < 130$	35
$130 \leqslant x < 140$	29
$140 \leqslant x < 150$	16
$150 \leqslant x < 160$	11
Total	110

(b)

Weight of egg in grams (x)	Number of eggs (f)
$20 \leqslant x < 25$	9
$25 \leqslant x < 30$	16
$30 \leqslant x < 35$	33
$35 \leqslant x < 40$	48
$40 \leqslant x < 45$	29
$45 \leqslant x < 50$	15
Total	150

(c)

Length of green bean in millimetres (x)	Frequency (f)
$60 \leqslant x < 80$	12
$80 \leqslant x < 100$	21
$100 \leqslant x < 120$	46
$120 \leqslant x < 140$	27
$140 \leqslant x < 160$	14
Total	120

(d)

Time to complete race in minutes (x)	Frequency (f)
$54 \leqslant x < 56$	1
$56 \leqslant x < 58$	4
$58 \leqslant x < 60$	11
$60 \leqslant x < 62$	6
$62 \leqslant x < 64$	2
$64 \leqslant x < 66$	1
Total	25

2 For each of these sets of data
   (i)   write down the modal class.   **(ii)**   calculate an estimate of the mean.

(a)

Height of shrub in metres (x)	Number of shrubs (f)
$0.3 \leqslant x < 0.6$	57
$0.6 \leqslant x < 0.9$	41
$0.9 \leqslant x < 1.2$	36
$1.2 \leqslant x < 1.5$	24
$1.5 \leqslant x < 1.8$	15

(b)

Weight of plum in grams (x)	Number of plums (f)
$20 \leqslant x < 30$	6
$30 \leqslant x < 40$	19
$40 \leqslant x < 50$	58
$50 \leqslant x < 60$	15
$60 \leqslant x < 70$	4

(c)

Length of journey in minutes (x)	Frequency (f)
$20 \leqslant x < 22$	6
$22 \leqslant x < 24$	20
$24 \leqslant x < 26$	38
$26 \leqslant x < 28$	47
$28 \leqslant x < 30$	16
$30 \leqslant x < 32$	3

(d)

Speed of car in miles per hour (x)	Frequency (f)
$25 \leqslant x < 30$	4
$30 \leqslant x < 35$	29
$35 \leqslant x < 40$	33
$40 \leqslant x < 45$	6
$45 \leqslant x < 50$	2
$50 \leqslant x < 55$	1

3   The table shows the monthly wages of the workers in an office.

Wages in £ (x)	$500 \leqslant x < 1000$	$1000 \leqslant x < 1500$	$1500 \leqslant x < 2000$	$2000 \leqslant x < 2500$
Frequency (f)	3	14	18	5

(a) What is the modal class?
(b) In which class is the median wage?
(c) Calculate an estimate of the mean wage.

4   The table shows the length, in seconds, of 100 calls made from a mobile phone.

Length of call in seconds (x)	$0 \leqslant x < 30$	$30 \leqslant x < 60$	$60 \leqslant x < 90$	$90 \leqslant x < 120$	$120 \leqslant x < 150$
Frequency (f)	51	25	13	7	4

Calculate an estimate of the mean length of a call.

5   The table shows the prices paid for greetings cards sold in one day by a card shop.

Calculate an estimate of the mean price, in pence, paid for a greetings card that day.

Price of greetings card in pence (x)	Frequency (f)
$75 \leqslant x < 100$	23
$100 \leqslant x < 125$	31
$125 \leqslant x < 150$	72
$150 \leqslant x < 175$	59
$175 \leqslant x < 200$	34
$200 \leqslant x < 225$	11
$225 \leqslant x < 250$	5

EXERCISE 10.1H

For each of these diagrams, find the area of the third square.

1

2

3

4

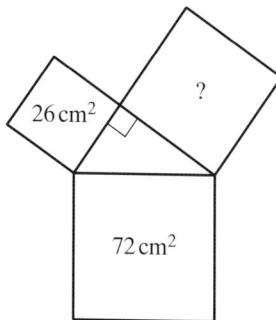

EXERCISE 10.2H

1 For each of these triangles, find the length marked x.
Where the answer is not exact, give your answer correct to 2 decimal places.

(a)

(b)

(c)

(d)

(e)

(f)

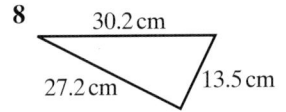

(g)

6.8 cm x 4.7 cm

(h)

x 18 cm 6.5 cm

7 8.4 cm 30 cm 28.8 cm

8 30.2 cm 27.2 cm 13.5 cm

2

Home Path Road (150 m) School Road (320 m)

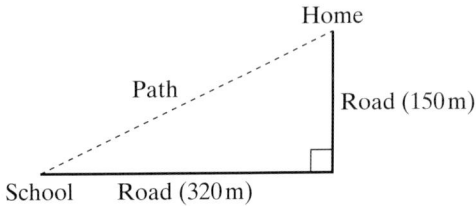

Ann can walk home from school
along two roads or along a path
across a field.
How much shorter is her journey if she takes
the path across the field?

3

20 cm 20 cm 10 cm

This network is made of wire.
What is the total length of wire?

EXERCISE 10.3H

Work out whether or not each of these triangles is
right-angled.
Show your working.

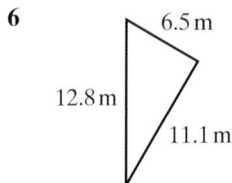

1 15 cm 20 cm 25 cm

2 7.3 m 4.2 m 8.7 m

3 12 cm 12 cm 15 cm

4 40.3 cm 15.5 cm 37.2 cm

5 8 m 11 m 7 m

6 6.5 m 12.8 m 11.1 m

EXERCISE 10.4H

1 Find the coordinates of the midpoint of each
of the lines in the diagram.

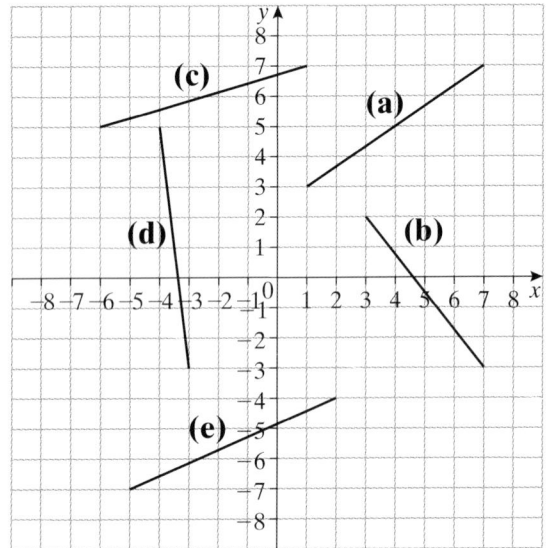

2 Find the coordinates of the midpoint of the
line joining each of these pairs of points.
Try to do them without plotting the points.
  (a) A(3, 7) and B(−5, 7)
  (b) C(2, 1) and D(8, 5)
  (c) E(3, 7) and F(8, 2)
  (d) G(1, 6) and H(9, 3)
  (e) I(−7, 1) and J(3, 6)
  (f) K(−5, −6) and L(−7, −3)

EXERCISE 10.5H

1 OABCDEFG is a cuboid.
F is the point (5, 7, 3).

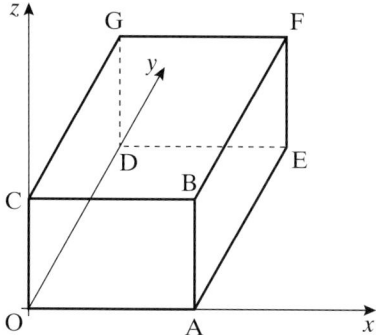

Write down the coordinates of
(a) point A. (b) point B.
(c) point C. (d) point D.
(e) point E. (f) point G.

2 OABCV is a pyramid with a rectangular base.
V is directly above the centre of the base, N.
OA = 8 units, AB = 10 units and
VN = 7 units.

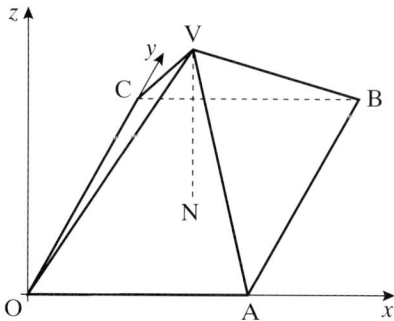

Write down the coordinates of
(a) point A. (b) point B.
(c) point C. (d) point N.
(e) point V.

3 OABCDEFG is a cuboid.
M is the midpoint of BF and N is the midpoint of GF.
OA = 6 units, OC = 5 units and
OD = 3 units.

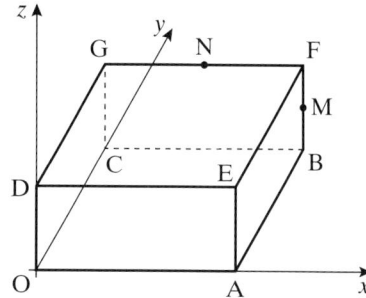

(a) Write down the coordinates of
   (i) point B.
   (ii) point F.
   (iii) point G.
   (iv) point M.
   (v) point N.
(b) (i) The point $(6, 2\frac{1}{2}, 0)$ is the midpoint of which edge?
   (ii) The point $(0, 2\frac{1}{2}, 1\frac{1}{2})$ is the centre of which face?

EXERCISE 11.1H

Work these out mentally. As far as possible, write down only the final answer.

1 (a) 8 + 25   (b) 13 + 49
   (c) 0.6 + 5.2   (d) 142 + 59
   (e) 187 + 25   (f) 5.8 + 12.6
   (g) 326 + 9.3   (h) 456 + 83
   (i) 1290 + 41   (j) 8600 + 570

2 (a) 12 − 5   (b) 36 − 9
   (c) 10 − 0.4   (d) 56 − 45
   (e) 82 − 39   (f) 141 − 27
   (g) 1.2 − 0.7   (h) 186 − 19
   (i) 307 − 81   (j) 1200 − 153

3 (a) 7 × 9   (b) 12 × 8
   (c) 22 × 7   (d) 11 × 12
   (e) 0.4 × 100   (f) 19 × 7
   (g) 48 × 5   (h) 25 × 8
   (i) 63 × 4   (j) 42 × 21

4 (a) 42 ÷ 7   (b) 72 ÷ 12
   (c) 1500 ÷ 3   (d) 184 ÷ 8
   (e) 240 ÷ 20   (f) 108 ÷ 18
   (g) 96 ÷ 16   (h) 16 ÷ 100
   (i) 24 ÷ 0.3   (j) 3.6 ÷ 0.9

5 (a) 7 + (−3)   (b) −2 + 6
   (c) −5 + 7   (d) −4 + (−8)
   (e) 4 + (−9)   (f) −5 − 1
   (g) 6 − (−2)   (h) 12 − (−8)
   (i) −5 − (−10)   (j) −7 − (−3)

6 (a) 12 × −2   (b) −4 × 5
   (c) −6 × −3   (d) −10 × −4
   (e) 4 × −100   (f) 10 ÷ −5
   (g) −6 ÷ 2   (h) −20 ÷ −4
   (i) −35 ÷ −5   (j) 32 ÷ −4

7 Write down the square of each of these numbers.
   (a) 7   (b) 9
   (c) 12   (d) 14
   (e) 1   (f) 30
   (g) 200   (h) 0.5
   (i) 0.8   (j) 0.2

8 Write down the square root of each of these numbers.
   (a) 36   (b) 4   (c) 64
   (d) 121   (e) 144

9 Write down the cube of each of these numbers.
   (a) 3   (b) 4   (c) 10
   (d) 50   (e) 0.2

10 A rectangle has sides 4.6 cm and 5.0 cm. Work out
   (a) the perimeter of the rectangle.
   (b) the area of the rectangle.

11 Ayeesha spends £17.81. How much change does she get from £50?

12 A square has area 121 cm². How long is its side?

13 Find 2% of £540.

14 Find two numbers the difference between which is 4 and the product 45.

15 Deepa has a 1 kg piece of cheese. She eats 432 g of it. How much is left?

1 Round each of these numbers to 1 significant figure.

(a) 14.3	**(b)** 38
(c) 6.54	**(d)** 308
(e) 1210	**(f)** 0.78
(g) 0.61	**(h)** 0.053
(i) 2413.5	**(j)** 0.0097

2 Round each of these numbers to 1 significant figure.

(a) 8.4	**(b)** 18.36
(c) 725	**(d)** 8032
(e) 98.3	**(f)** 0.71
(g) 0.0052	**(h)** 0.019
(i) 407.511	**(j)** 23 095

3 Round each of these numbers to 2 significant figures.

(a) 28.7	**(b)** 149.3
(c) 7832	**(d)** 46 820
(e) 21.36	**(f)** 0.194
(g) 0.0489	**(h)** 0.003 61
(i) 0.0508	**(j)** 0.904

4 Round each of these numbers to 3 significant figures.

(a) 7.385	**(b)** 24.81
(c) 28 462	**(d)** 308.61
(e) 16 418	**(f)** 3.917
(g) 60.72	**(h)** 0.9135
(i) 0.004 162	**(j)** 2.236 06

For questions **5** to **12**, round the numbers in your calculations to 1 significant figure.
Show your working.

5 Ashad bought 24 chocolate bars at 32p each. Estimate how much he spent.

6 Ramy earns £382 per week. Estimate his earnings in a year.

7 Mary drove 215 miles in 3 hours 48 minutes. Estimate her average speed.

8 A rectangle has length 9.2 cm and area 44.16 cm^2. Estimate its width.

9 A new computer is priced at £595 excluding VAT.
VAT at 17.5% must be paid on it.
Estimate the amount of VAT to be paid.

10 A square paving slab has area 6000 cm^2. Estimate the length of a side of the slab.

11 A circle has radius 4.3 cm. Estimate its area.

12 Estimate the answers to these calculations.

(a) 71×58	**(b)** $\sqrt{46}$
(c) $\dfrac{5987}{5.1}$	**(d)** 19.1^2
(e) 62.7×8316	**(f)** $\dfrac{5.72}{19.3}$
(g) $\dfrac{32}{49.4}$	**(h)** 8152×37
(i) $\dfrac{935 \times 41}{8.5}$	**(j)** $\dfrac{673 \times 0.76}{3.6 \times 2.38}$

■ **EXERCISE 11.3H** 🔵pt

Give your answers to these questions as simply as possible.
Leave π in your answers where appropriate.

1 **(a)** $2 \times 6 \times \pi$ **(b)** $\pi \times 7^2$
    (c) $\pi \times 12^2$ **(d)** $2 \times 3.8 \times \pi$
    (e) $\pi \times 11^2$

2 **(a)** $14\pi + 5\pi$
    (b) $\pi \times 3^2 + \pi \times 6^2$
    (c) $\pi \times 8^2 - \pi \times 4^2$
    (d) $3 \times 42\pi$
    (e) $\dfrac{36\pi}{4\pi}$

3 Find the circumference of a circle with radius 15 cm.

4 The areas of two circles are in the ratio $36\pi : 16\pi$. Simplify this ratio.

5 A circular piece of card has radius 12 cm. A square piece is removed, with side 5 cm. Find the area that is left.

EXERCISE 11.4H

1 Work out these.
   (a) 0.06×600 (b) 0.03×0.3
   (c) 0.9×0.04 (d) $(0.05)^2$
   (e) $(0.3)^2$ (f) 500×800
   (g) 30×5000 (h) 5.1×300
   (i) 20.3×2000 (j) 1.82×5000

2 Work out these.
   (a) $300 \div 20$ (b) $60 \div 2000$
   (c) $3.6 \div 20$ (d) $1.4 \div 0.2$
   (e) $2.4 \div 3000$ (f) $2.4 \div 0.03$
   (g) $0.08 \div 0.004$ (h) $5 \div 0.02$
   (i) $400 \div 0.08$ (j) $60 \div 0.15$

3 Given that $4.5 \times 16.8 = 75.6$, work out these.
   (a) 45×1680 (b) $75.6 \div 168$
   (c) $7560 \div 45$ (d) 0.168×0.045
   (e) $756 \div 0.168$

4 Given that $702 \div 39 = 18$, work out these.
   (a) $70\,200 \div 39$ (b) $70.2 \div 3.9$
   (c) 180×39 (d) $7.02 \div 18$
   (e) 1.8×3.9

5 Given that $348 \times 216 = 75\,168$, work out these.
   (a) $751\,680 \div 216$ (b) $34\,800 \times 2160$
   (c) 3.48×21.6 (d) $751.68 \div 34.8$
   (e) 0.348×2160

12 → FORMULAE 1

EXERCISE 12.1H

1 Pearl is a child minder. She charges £3.50 an hour.
She looks after Mrs Khan's child for 6 hours. How much does she charge?

2 It costs £60 plus £1 a mile to hire a coach.
(a) How much does it cost to hire a coach to go
   (i)   80 miles?   (ii) 150 miles?
(b) Write a formula for the cost, £C, of hiring a coach to go n miles.

3 The cost of booking a room for a meeting is £80 plus £20 an hour.
(a) How much does it cost to hire the room for
   (i)   5 hours?   (ii) 8 hours?
(b) Write a formula for the cost, £C, of hiring the room for h hours.

4 The perimeter of a rectangle is twice the length plus twice the width.
(a) What is the perimeter of a rectangle with length 5 cm and width 3.5 cm?
(b) Write a formula for the perimeter, P, of a rectangle with length x and width y.

5 To find the volume of a pyramid, multiply the area of the base by the height and divide by 3.
(a) What is the volume of a pyramid with base area 12 cm² and height 7 cm?
(b) Write a formula for the volume, V, of a pyramid with base area A and height h

6 To find the time it takes to type a document, divide the number of words in the document by the number of words typed per minute.
(a) How long does it take Liz to type a document 560 words long if she typed 80 words a minute?

(b) Write a formula for the time, T, to type a document w words long if the typist types r words a minute.

7 The time, t, for a journey is the distance, d, divided by the speed, s.
(a) Write a formula for this.
(b) Steve travelled 175 miles at a speed of 50 mph. How long did the journey take?

8 The circumference of a circle is given by the formula:

   $C = \pi \times D$, where D is the diameter of the circle.

Find the circumference of a circle with diameter 8.5 cm. Use $\pi = 3.14$.

9 At Carterknowle toddlers group the charge is £1 per carer and 50p for every toddler they bring.
(a) Tracey brings three toddlers to the group. How much does she pay?
(b) Fran brings n toddlers to the group. Write down an equation for the amount, £A, she has to pay.

10 (a) For the formula $A = b - c$, find A when $b = 6$ and $c = 3.5$.
(b) For the formula $B = 2a - b$, find B when $a = 6$ and $b = 5$.
(c) For the formula $C = 2a - b + 3c$, find C when $a = 3.5$, $b = 2.6$ and $c = 1.2$.
(d) For the formula $D = 3b^2$, find D when $b = 2$.
(e) For the formula $E = ab - cd$, find E when $a = 12.5$, $b = 6$, $c = 3.5$ and $d = 8$.
(f) For the formula $F = \dfrac{a - b}{5}$, find F when $a = 6$ and $b = 3.5$.

EXERCISE 12.2H pt

1 Rearrange each of these formulae to make the letter in brackets the subject.

(a) $a = b + c$ (b)

(b) $a = 3x - y$ (x)

(c) $a = b + ct$ (t)

(d) $F = 2(q + p)$ (q)

(e) $x = 2y - 3z$ (y)

(f) $P = \dfrac{3 + 4n}{5}$ (n)

2 The formula for the circumference of a circle is $C = \pi d$.
Rearrange the formula to make d the subject.

3 Rearrange the formula $A = \dfrac{3ab}{2n}$ to make

(a) a the subject.

(b) n the subject.

4 The formula for finding the perimeter of a rectangle is $P = 2(a + b)$, where P is the perimeter, a is the length and b is the width of the rectangle.
Rearrange the formula to make a the subject.

5 The formula $y = mx + c$ is the equation of a straight line.
Rearrange it to find m in terms of x, y and c.

6 The surface area of a sphere is given by the formula $A = 4\pi r^2$.
Rearrange the formula to make r the subject.

7 The formula for the volume of this prism is $V = \dfrac{\pi r^2 h}{4}$.

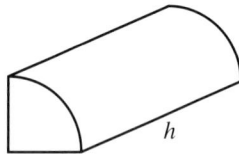

(a) Find V when $r = 2.5$ and $h = 7$.

(b) (i) Rearrange the formula to make r the subject.

(ii) Find r when $V = 100$ and $h = 10$.

8 The formula for the surface area of a closed cylinder is $S = 2\pi r(r + h)$.
Rearrange the formula to make h the subject.

EXERCISE 12.3H

1 Calculate the value of $x^3 + x$ when
(a) $x = 2$.   (b) $x = 3$.   (c) $x = 2.5$.

2 (a) Calculate the value of $x^3 - x$ when
(i) $x = 4$.   (ii) $x = 5$.
(iii) $x = 4.6$.   (iv) $x = 4.7$.
(v) $x = 4.65$.

(b) Using your answers to part (a), give the solution of $x^3 - x = 94$, to 1 decimal place.

Use trial and improvement for questions 3 to 10. Show your trials.

3 Find a solution, between $x = 2$ and $x = 3$, to the equation $x^3 = 11$.
Give your answer correct to 1 decimal place.

4 (a) Show that a solution to the equation $x^3 + 3x = 30$ lies between $x = 2$ and $x = 3$.

(b) Find the solution correct to 1 decimal place.

5 (a) Show that a solution to the equation $x^3 - 2x = 70$ lies between $x = 4$ and $x = 5$.

(b) Find the solution correct to 1 decimal place.

6 Find a solution to the equation $x^3 + 4x = 100$.
Give your answer correct to 1 decimal place.

7 Find a solution to the equation $x^3 + x = 60$.
Give your answer correct to 2 decimal places.

8 Find a solution to the equation $x^3 - x^2 = 40$.
Give your answer correct to 2 decimal places.

9 A number, x, added to the square of that number is equal to 1000.
(a) Write this as a formula.
(b) Find the number correct to 1 decimal place.

10 The cube of a number minus the number is equal to 600.
Find the number correct to 2 decimal places.

EXERCISE 13.1H

1 Draw a pair of axes and label them
−2 to 4 for *x* and *y*.
  (a) Draw a triangle with vertices at
  (1, 1), (1, 3) and (0, 3). Label it A.
  (b) Reflect triangle A in the line *x* = 2.
  Label it B.
  (c) Reflect triangle A in the line *y* = *x*.
  Label it C.
  (d) Reflect triangle A in the line *y* = 2.
  Label it D.

2 Draw a pair of axes and label them
−3 to 3 for *x* and *y*.
  (a) Draw a triangle with vertices at (−1, 1),
  (−1, 3) and (−2, 3). Label it A.
  (b) Reflect triangle A in the line $x = \frac{1}{2}$.
  Label it B.
  (c) Reflect triangle A in the line *y* = *x*.
  Label it C.
  (d) Reflect triangle A in the line
  y = −*x*. Label it D.

3 For each part
  • copy the diagram.
  • reflect the shape in the mirror line.
  (a)   **(b)**

  (c)

4 Describe fully the single transformation that maps
  (a) shape A on to shape B.
  (b) shape A on to shape C.
  (c) shape B on to shape D.

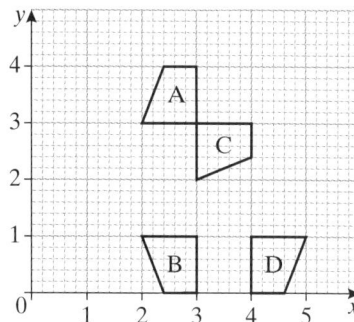

5 Describe fully the single transformation that maps
  (a) triangle A on to triangle B.
  (b) triangle A on to triangle C.
  (c) triangle E on to triangle F.

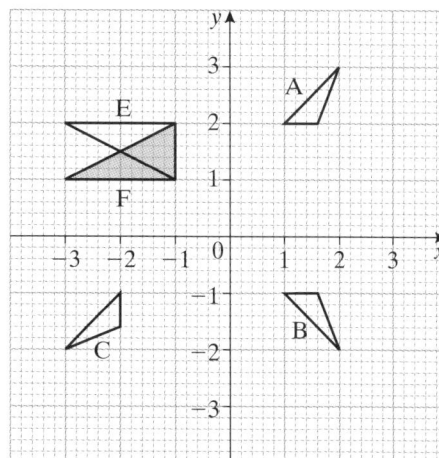

EXERCISE 13.2H

1   Copy the diagram.
    (a) Rotate trapezium A through 180° about
        the origin. Label it B.
    (b) Rotate trapezium A through 90°
        clockwise about the point (0, 1). Label it C.
    (c) Rotate trapezium A through 90°
        anticlockwise about the point
        (−1, 1). Label it D.

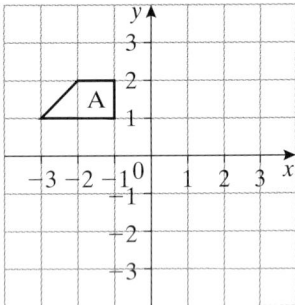

2   Copy the diagram.
    (a) Rotate flag A through 90° clockwise
        about the origin. Label it B.
    (b) Rotate flag A through 90° anticlockwise
        about the point (1, −1). Label it C.
    (c) Rotate flag A through 180° about the
        point (0, −1) Label it D.

3   Draw a pair of axes and label them −4 to 4 for
    x and y.
    (a) Draw a triangle with vertices
        (1, 1), (2, 1) and (2, 3). Label it A.
    (b) Rotate triangle A through 90°
        anticlockwise about the origin. Label it B.
    (c) Rotate triangle A through 180° about the
        point (2, 1). Label it C.
    (d) Rotate triangle A through 90° clockwise
        about the point (−2, 1). Label it D.

4   Copy the diagram.
    Rotate the triangle
    through 180° about the
    point C.

5   Copy the diagram.
    Rotate the shape through
    90° clockwise about the
    point O.

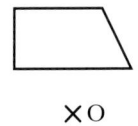

6   Copy the diagram.
    Rotate the flag through
    150° clockwise about the
    point A.

7   Describe fully the single transformation that
    maps
    (a) trapezium A on to trapezium B.
    (b) trapezium A on to trapezium C.
    (c) trapezium A on to trapezium D.

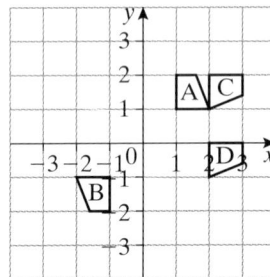

8   Describe fully the transformation that maps
    (a) flag A on to flag B.
    (b) flag A on to flag C.
    (c) flag A on to flag D.

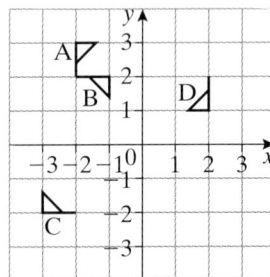

9   Describe fully the
    single transformation
    that maps triangle A on
    to triangle B.

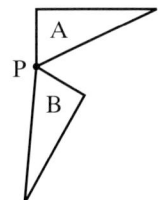

10 Describe fully the single transformation that maps
   (a) triangle A on to triangle B.
   (b) triangle A on to triangle C.
   (c) triangle A on to triangle D.
   (d) triangle A on to triangle E.
   (e) triangle A on to triangle F.
   Hint: Some of these transformations are reflections.

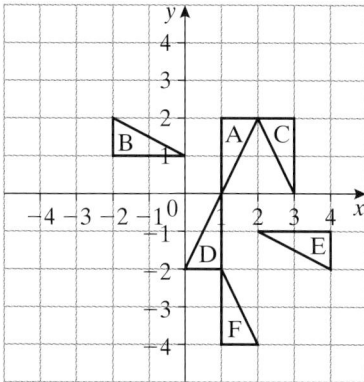

EXERCISE 13.3H

1 Draw a pair of axes and label them −2 to 6 for x and y.
   (a) Draw a triangle with vertices at (1, 1), (1, 2), and (4, 1). Label it A.
   (b) Translate A by vector $\begin{pmatrix} 1 \\ 3 \end{pmatrix}$. Label it B.
   (c) Translate A by vector $\begin{pmatrix} -3 \\ 4 \end{pmatrix}$. Label it C.
   (d) Translate A by vector $\begin{pmatrix} -2 \\ -3 \end{pmatrix}$. Label it D.

2 Draw a pair of axes and label them −3 to 5 for x and y.
   (a) Draw a triangle with vertices at (2, 1), (2, 3) and (3, 1). Label it A.
   (b) Translate A by vector $\begin{pmatrix} 2 \\ 1 \end{pmatrix}$. Label it B.
   (c) Translate A by vector $\begin{pmatrix} -5 \\ -3 \end{pmatrix}$. Label it C.
   (d) Translate A by vector $\begin{pmatrix} 2 \\ -4 \end{pmatrix}$. Label it D.

3 Describe the single transformation that maps
   (a) triangle A on to triangle B.
   (b) triangle A on to triangle C.
   (c) triangle A on to triangle D.
   (d) triangle B on to triangle D.

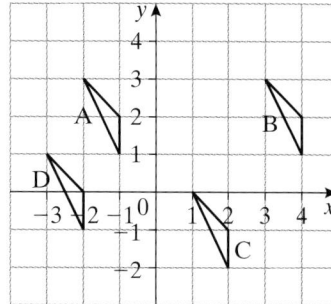

4 Describe the single transformation that maps
   (a) shape A on to shape B.
   (b) shape A on to shape C.
   (c) shape A on to shape D.
   (d) shape D on to shape E.
   (e) shape A on to shape F.
   (f) shape E on to shape G.
   (g) shape B on to shape H.
   (h) shape H on to shape F.
   Hint: Not all the transformations are translations.

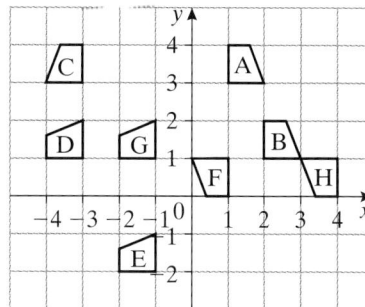

EXERCISE 13.4H

1 Draw a pair of axes and label them 0 to 6 for both x and y.
   (a) Draw a triangle with vertices at (0, 6), (3, 6) and (3, 3). Label it A.
   (b) Enlarge triangle A by scale factor $\frac{1}{3}$, with the origin as the centre of enlargement. Label it B.
   (c) Describe fully the single transformation that maps triangle B on to triangle A.

2 Draw a pair of axes and label them 0 to 6 for both x and y.
  (a) Draw a triangle with vertices at (5, 2), (5, 6) and (3, 6). Label it A.
  (b) Enlarge triangle A by scale factor $\frac{1}{2}$, with centre of enlargement (3, 2). Label it B.
  (c) Describe fully the single transformation that maps triangle B on to triangle A.

3 Draw a pair of axes and label them 0 to 8 for both x and y.
  (a) Draw a triangle with vertices at (2, 1), (2, 3), (3, 2). Label it A.
  (b) Enlarge triangle A by scale factor $2\frac{1}{2}$, with the origin as the centre of enlargement. Label it B.
  (c) Describe fully the single transformation that maps triangle B on to triangle A.

4 Draw a pair of axes and label them 0 to 7 for both x and y.
  (a) Draw a trapezium with vertices at (1, 2), (1, 3), (2, 3) and (3, 2). Label it A.
  (b) Enlarge triangle A by scale factor 3, with centre of enlargement (1, 2). Label it B.
  (c) Describe fully the single transformation that maps triangle B on to triangle A.

5 Describe fully the single transformation that maps
  (a) triangle A on to triangle B.
  (b) triangle B on to triangle A.
  (c) triangle A on to triangle C.
  (d) triangle C on to triangle A.

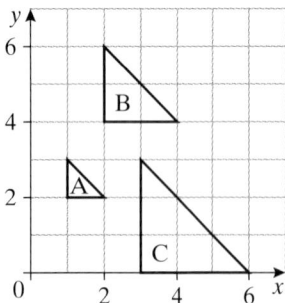

Hint: in questions **6**, **7** and **8**, not all the transformations are enlargements.

6 Describe fully the single transformation that maps
  (a) flag A on to flag B.
  (b) flag B on to flag C.
  (c) flag B on to flag D.
  (d) flag B on to flag E.
  (e) flag F on to flag G.

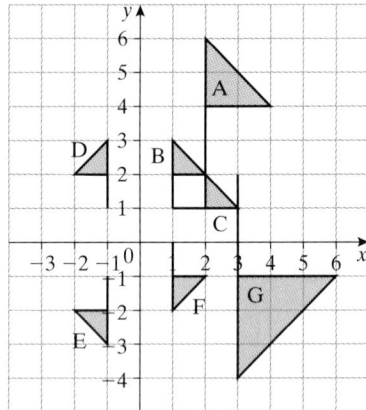

7 Draw a pair of axes and label them -4 to 4 for x and y.
  (a) Draw a triangle with vertices at (2, 1), (2, 2) and (4, 2). Label it A.
  (b) Reflect A in the line $y = 0$. Label it B.
  (c) Reflect A in the line $x = 1$. Label it C.
  (d) Rotate B by 90° about the origin. Label it D.
  (e) Enlarge A by scale factor $\frac{1}{2}$, with the origin as the centre of enlargement. Label it E.

8 Copy the diagram.
  (a) Rotate shape A through 90° clockwise about the origin. Label it B.
  (b) Rotate shape A through 180° about (2, 2). Label it C.
  (c) Enlarge shape A by scale factor $\frac{1}{2}$, with centre of enlargement $(-2, 0)$. Label it E.

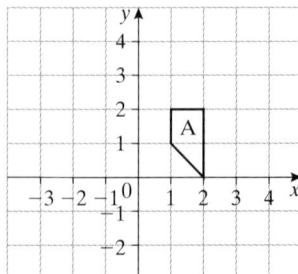

EXERCISE 14.1H

1   The probability that Stacey will go to bed late tonight is 0.2.
    What is the probability that Stacey will not go to bed late tonight?

2   The probability that I will throw a six with a dice is $\frac{1}{6}$.
    What is the probability that I will not throw a six?

3   The probability that it will snow on Christmas Day is 0.15.
    What is the probability that it will not snow on Christmas Day?

4   The probability that someone chosen at random is left-handed is $\frac{3}{10}$.
    What is the probability that they will be right-handed?

5   The probability that United will lose their next game is 0.08.
    What is the probability that United will not lose their next game?

6   The probability that Ian will eat crisps one day is $\frac{17}{31}$.
    What is the probability that he will not eat crisps?

EXERCISE 14.2H

1   A shop has brown, white and wholemeal bread for sale.
    The probability that someone will choose brown bread is 0.4 and the probability that they will choose white bread is 0.5.
    What is the probability of someone choosing wholemeal bread?

2   A football coach is choosing a striker for the next game.
    He has three players to choose from; Wayne, Michael and Alan.
    The probability that he will choose Wayne is $\frac{5}{19}$ and the probability that he will choose Michael is $\frac{7}{19}$.
    What is the probability that he will choose Alan?

3   A bag contains red, white and blue counters. Jill chooses a counter at random.
    The probability that she chooses a red counter is 0.4 and the probability that she chooses a blue counter is 0.15.
    What is the probability that she chooses a white counter?

4   Elaine goes to town by car, bus, taxi or bike. The probability that she uses her car is $\frac{12}{31}$, the probability that she catches the bus is $\frac{2}{31}$ and the probability that she takes a taxi is $\frac{13}{31}$.
    What is the probability that she rides her bike into town?

5   A biased five-sided spinner is numbered 1 to 5. The table shows the probability of obtaining some of the scores when it is spun.

Score	1	2	3	4	5
Probability	0.37	0.1	0.14		0.22

    What is the probability of getting 4?

6   A cash bag contains only £20, £10 and £5 notes. One note is chosen from the bag at random.
    There is a probability of $\frac{3}{4}$ that it is a £5 note and a probability of $\frac{3}{20}$ that it is a £10 note.
    What is the probability that it is a £20 note?

EXERCISE 14.3H

1   The probability that United will lose their next game is 0.2.
    How many games would you expect them to lose in a season of 40 games?

2   The probability that it will be rainy on any day in June is $\frac{2}{15}$.
    On how many of June's 30 days would you expect it to be rainy?

3   The probability that an eighteen-year-old driver will have an accident is 0.15.
    There are 80 eighteen-year-old drivers in a school.
    How many of them might be expected to have an accident?

4   When Phil is playing chess, the probability that he wins is $\frac{17}{20}$.
    In a competition, Phil plays 10 games. How many of them might you expect him to win?

5   An ordinary six-sided dice is thrown 90 times.
    How many times might you expect to get
    (a) a 4?   (b) an odd number?

6   A box contains twelve yellow balls, three blue balls and five green balls.
    A ball is chosen at random and its colour noted.
    The ball is then replaced. This is done 400 times.
    How many of each colour might you expect to get?

EXERCISE 14.4H

1   Pete rolls a dice 200 times and records the number of times each score appears.

Score	1	2	3	4	5	6
Frequency	29	34	35	32	34	36

    (a) Work out the relative frequency of each of the scores.
        Give your answers to 2 decimal places.
    (b) Do you think that Pete's dice is fair? Give a reason for your answer.

2   Rory kept a record of his favourite football team's results.
       Win: 32   Draw: 11   Lose: 7
    (a) Calculate the relative frequency of each of the three outcomes.
    (b) Are your answers to part (a) good estimates of the probability of the outcome of their next match?
        Give a reason for your answer.

3   In a survey, 600 people were asked which flavour of crisps they preferred.
    The results are shown in the table.

Flavour	Frequency
Plain	166
Salt & Vinegar	130
Cheese & Onion	228
Other	76

    (a) Work out the relative frequency for each flavour.
        Give your answers to 2 decimal places.
    (b) Explain why it is reasonable to use these figures to estimate the probability of the flavour of crisps that the next person to be asked will prefer.

4   The owner of a petrol station notices that in one day 287 out of 340 people filling their car with petrol spent over £20.
    Use these figures to estimate the probability that the next customer will spend
    (a) over £20.
    (b) £20 or less.

5 Jasmine made a spinner numbered 1, 2, 3, 4 and 5.
She tested the spinner to see if it was fair.
The results are shown below.

Score	1	2	3	4	5
Frequency	46	108	203	197	96

(a) Work out the relative frequency of each of the scores.
Give your answers to 2 decimal places.
(b) Do you think that the spinner is fair?
Give a reason for your answer.

6 A box contains yellow, green, white and blue counters.
A counter is chosen from the box and its colour noted. The counter is then replaced in the box.
The table below gives information about the colour of counter picked.

Colour	Relative frequency
Yellow	0.4
Green	0.3
White	0.225
Blue	0.075

(a) There are 80 counters altogether in the bag. How many do you think there are of each colour?
(b) What other information is needed before you can be sure that your answers to part (a) are accurate?

EXERCISE 15.1H

1 Draw the graph of $y = 3x$ for values of x from -3 to 3.

2 Draw the graph of $y = x + 2$ for values of x from -4 to 2.

3 Draw the graph of $y = 4x + 2$ for values of x from -3 to 3.

4 Draw the graph of $y = 2x - 5$ for values of x from -1 to 5.

5 Draw the graph of $y = -2x - 4$ for values of x from -4 to 2.

EXERCISE 15.2H

1 Draw the graph of $3y = 2x + 6$ for values of x from -3 to 3.

2 Draw the graph of $2x + 5y = 10$.

3 Draw the graph of $3x + 2y = 15$.

4 Draw the graph of $2y = 5x - 8$, for values of x from -2 to 4.

5 Draw the graph of $3x + 4y = 24$.

EXERCISE 15.3H

1 Simon had a bath. The graph shows the volume (V gallons) of the water in the bath after t minutes.

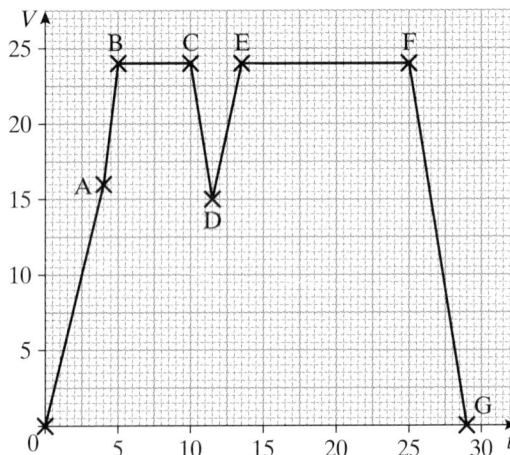

(a) How many gallons of water are in the bath at A?

(b) Simon got in the bath at B and out at F. How long was he in the bath?

(c) Between O and A, the hot tap is on. How many gallons of water per minute came from the hot tap?

(d) Between A and B, both taps are on. What is the rate of flow of both taps together?
Give your answer in gallons/minute.

(e) Describe what happened between C and E.

(f) At what rate did the bath empty? Give your answer in gallons/minute.

2 A printer's charge for printing programmes is worked out as follows.

A fixed charge of £*a*
+
x pence per programme for the first 1000 programmes
+
80 pence per programme for each programme over 1000

The graph below shows the total charge for printing up to 1000 programmes.

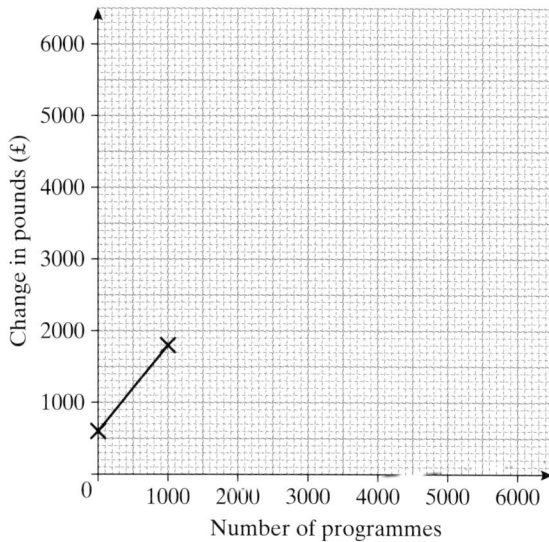

(a) What is the fixed charge, £*a*?
(b) Calculate *x*, the charge per programme for the first 1000 programmes.
(c) Copy the graph and add a line segment to show the charges for 1000 to 6000 programmes.
(d) What is the total charge for 3500 programmes?
(e) What is the average cost per programme for 3500 programmes?

3 The graph shows a train journey.

(a) How long did the train journey take?
(b) How far was the train journey?
(c) How far from the start was the first station?
(d) How long did the train stop at the first station?
(e) When was the train travelling fastest?

4 Water is poured into each of these vessels at a constant rate until they are full.

(a) (b)

(c) (d)

These graphs show depth of water (*d*) against time (*t*).
Choose the most suitable graph for each vessel.

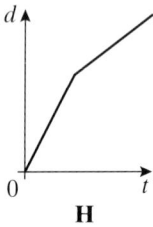

A

B

C

D

E

F

G

H

5 A mobile phone company offers its customers the choice of two price plans.

	Plan A	Plan B
Monthly subscription	£10.00	£*s*
Free talk time per month	60 minutes	100 minutes
Cost per minute over the free talk time	*a* pence	35 pence

The graph shows the charges for Plan A and the charges for Plan B up to 100 minutes.

(a) Find the monthly subscription for Plan B (£*s*).
(b) Shamir uses price plan B. How much does it cost him if he uses the phone for 250 minutes per month?
(c) Copy the graph and add a line to show the cost for Plan B for 100 to 250 minutes.
(d) For how many minutes is the cost the same in both price plans?
(e) Which price plan is the cheaper when the time for calls is 220 minutes? By how much?

6 The table shows the cost of sending parcels.

Maximum weight	Cost
10 kg	£13.85
11 kg	£14.60
12 kg	£15.35
13 kg	£16.10
14 kg	£16.85
15 kg	£17.60

The graph shows the information in the first row in the table.

(a) What is the cost of sending a parcel weighing
  (i) 9.6 kg?
  (ii) 10 kg?
  (iii) 10.1 kg?

(b) (i) What is the meaning of the dot at the right of the line?
  (ii) What is the meaning of the circle at the left of the line?
(c) Copy the graph and add lines to show the cost for parcels weighing up to 15 kg.
(d) Hazel posted one parcel weighing 8.4 kg and another weighing 12.8 kg. What was the total cost?

7 15 000 people attended a football match.
The gates opened 60 minutes before kick off.
The match lasted 90 minutes with a break of 15 minutes at half-time.
200 people left at half-time.
Some people started leaving 10 minutes before the final whistle, but most stayed to the end.
The grounds were cleared 30 minutes after the final whistle.
Copy the axes below and sketch a graph of the number of people in the grounds against the time after the gates opened.

EXERCISE 15.4H 🅟

1   Which of these functions are quadratic? For each of the functions that is quadratic, state whether the graph is ∪-shaped or ∩-shaped.

(a) $y = x^2 + 7$ (b) $y = 2x^3 + x^2 - 4$ (c) $y = x^2 + x - 7$ (d) $y = x(6 - x)$

(e) $y = \dfrac{5}{x^2}$ (f) $y = x(x^2 + 1)$ (g) $y = 2x(x + 2)$ (h) $y = 5 + 3x - x^2$

2   (a) Copy and complete the table of values for $y = 2x^2$.

x	-3	-2	-1	0	1	2	3
x^2	9					4	
$y = 2x^2$	18					8	

(b) Plot the graph of $y = 2x^2$.
    Use a scale of 2 cm to 1 unit on the x-axis and 1 cm to 1 unit on the y-axis.
(c) Use your graph to
    (i)   find the value of y when $x = -1.8$.
    (ii) solve $2x^2 = 12$.

3   (a) Copy and complete the table of values for $y = x^2 + x$.

x	-4	-3	-2	-1	0	1	2	3
x^2			4					9
$y = x^2 + x$								12

(b) Plot the graph of $y = x^2 + x$.
    Use a scale of 2 cm to 1 unit on the x-axis and 1 cm to 1 unit on the y-axis.
(c) Use your graph to
    (i)   find the value of y when $x = 1.6$.
    (ii) solve $x^2 + x = 8$.

4   (a) Copy and complete the table of values for $y = x^2 - x + 2$.

x	-3	-2	-1	0	1	2	3	4
x^2		4						16
$-x$		2						-4
2		2						2
$y = x^2 - x + 2$		8						14

(b) Plot the graph of $y = x^2 - x + 2$.
    Use a scale of 2 cm to 1 unit on the x-axis and 1 cm to 1 unit on the y-axis.
(c) Use your graph to
    (i)   find the value of y when $x = 0.7$
    (ii) solve $x^2 - x + 2 = 6$.

5 **(a)** Copy and complete the table of values for $y = x^2 + 2x - 5$.

x	-5	-4	-3	-2	-1	0	1	2	3
x^2				4					9
$2x$				-4					6
-5				-5					-5
$y = x^2 + 2x - 5$				-5					10

(b) Plot the graph of $y = x^2 + 2x - 5$.
Use a scale of 2 cm to 1 unit on the x-axis and 1 cm to 1 unit on the y-axis.
(c) Use your graph to
    (i) find the value of y when $x = -1.4$.   **(ii)** solve $x^2 + 2x - 5 = 0$.

6 **(a)** Copy and complete the table of values for $y = 8 - x^2$.

x	-3	-2	-1	0	1	2	3
8				8			8
$-x^2$				0			-9
$y = 8 - x^2$				8			-1

(b) Plot the graph of $y = 8 - x^2$.
Use a scale of 2 cm to 1 unit on the x-axis and 1 cm to 1 unit on the y-axis.
(c) Use your graph to
    (i) find the value of y when $x = 0.5$.   **(ii)** solve $8 - x^2 = -2$.

7 **(a)** Copy and complete the table of values for $y = (x - 2)(x + 1)$.

x	-3	-2	-1	0	1	2	3	4
$x - 2$		-4					1	
$x + 1$		-1					4	
$y = (x - 2)(x + 1)$		4					4	

(b) Plot the graph of $y = (x - 2)(x + 1)$.
Use a scale of 2 cm to 1 unit on the x-axis and 1 cm to 1 unit on the y-axis.
(c) Use your graph to
    (i) find the minimum value of y.   **(ii)** solve $(x - 2)(x + 1) = 2.5$.

8 **(a)** Make a table of values for $y = x^2 - 3x + 2$. Choose values of x from -2 to 5.
(b) Plot the graph of $y = x^2 - 3x + 2$.
Use a scale of 2 cm to 1 unit on the x-axis and 1 cm to 1 unit on the y-axis.
(c) Use your graph to solve
    (i) $x^2 - 3x + 2 = 1$.   **(ii)** $x^2 - 3x + 2 = 10$.

16 → MEASURES

EXERCISE 16.1H

1 Change these units.
- **(a)** 25 cm to mm
- **(b)** 24 m to cm
- **(c)** 1.36 cm to mm
- **(d)** 15.1 cm to mm
- **(e)** 0.235 m to mm

2 Change these units.
- **(a)** 2 m^2 to cm^2
- **(b)** 3 cm^2 to mm^2
- **(c)** 1.12 m^2 to cm^2
- **(d)** 0.05 cm^2 to mm^2
- **(e)** 2 m^2 to mm^2

3 Change these units.
- **(a)** 8000 mm^2 to cm^2
- **(b)** 84 000 mm^2 to cm^2
- **(c)** 2 000 000 cm^2 to m^2
- **(d)** 18 000 000 cm^2 to m^2
- **(e)** 64 000 cm^2 to m^2

4 Change these units.
- **(a)** 32 cm^3 to mm^3
- **(b)** 24 m^3 to cm^3
- **(c)** 5.2 cm^3 to mm^3
- **(d)** 0.42 m^3 to cm^3
- **(e)** 0.02 cm^3 to mm^3

5 Change these units.
- **(a)** 5 200 000 cm^3 to m^3
- **(b)** 270 000 mm^3 to cm^3
- **(c)** 210 cm^3 to m^3
- **(d)** 8.4 m^3 to mm^3
- **(e)** 170 mm^3 to cm^3

6 Change these units.
- **(a)** 36 litres to cm^3
- **(b)** 6300 ml to litres
- **(c)** 1.4 litres to ml
- **(d)** 61 ml to litres
- **(e)** 5400 cm^3 to litres.

EXERCISE 16.2H

1 Copy and complete each of these statements.
- **(a)** A length given as 4.3 cm, to 1 decimal place, is between .... cm and .... cm.
- **(b)** A capacity given as 463 ml, to the nearest millilitre, is between .... ml and .... ml
- **(c)** A time given as 10.5 seconds, to the nearest tenth of a second, is between .... seconds and .... seconds.
- **(d)** A mass given as 78 kg, to the nearest kilogram, is between .... kg and .... kg.
- **(e)** An area given as 5.5 m^2, to 1 decimal place, is between .... m^2 and .... m^2.

2 The number of people attending a football match was given as 24 000 to the nearest thousand.
What was the least number of people that could have been at the match?

3 Kerry measures her height as 142 cm to the nearest centimetre.
Write down the two values between which her height must lie.

4 The height of a desk is stated as 75.0 cm to 1 decimal place.
Write down the two values between which its height must lie.

5 Rashid measures the thickness of a piece of plywood as 7.83 mm, to 2 decimal places.
Write down the smallest and greatest thickness it could be.

6 The sides of this triangle are given in centimetres, correct to 1 decimal place.

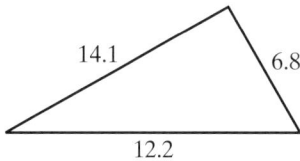

(a) Write down the shortest and longest possible length of each side.

(b) Write down the shortest and longest possible length of the perimeter.

7 John has two pieces of string.
He measures them as 125 mm and 182 mm, to the nearest millimetre.
He puts the two pieces end to end.
What is the shortest and longest that their combined lengths can be?

8 Mel and Mary both buy some apples. Mel buys 3.5 kg and Mary buys 4.2 kg.
Both weights are correct to the nearest tenth of a kilogram.

(a) What is the smallest possible difference between the amounts they have bought?

(b) What is the largest possible difference between the amounts they have bought?

EXERCISE 16.3H

1 Rewrite each of these statements using sensible values for the measurements.

(a) My mass is 78.32 kg

(b) It takes Katriona 16 minutes and 15.6 seconds to walk to school.

(c) The distance to London from Sheffield is 161.64 miles.

(d) The length of our classroom is 5 metres 14 cm 3 mm.

(e) My water jug hold 3.02 litres.

2 Give your answer to each of these questions to a sensible degree of accuracy.

(a) Estimate the length of this line.

(b) Estimate the size of this angle.

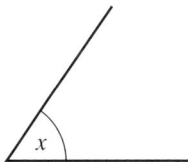

(c) A rectangle is 2.3 cm long and 4.5 cm wide. Find the area of the rectangle.

(d) The diameter of a circle is 8 cm. Work out the circumference.

(e) The volume of a cube is 7 cm³. Find the length of an edge.

(f) An angle in a pie chart is found by working out $\frac{4}{7} \times 360°$. Find the angle.

(g) A coach travels 73 miles at an average speed of 33 mph. How long does it take?

(h) Six friends share £14 between them. How much does each one get?

EXERCISE 16.4H pt

1 A boat travels a distance of 24 km in a time of 3 hours.
   Calculate its average speed.

2 A car covers 197 miles on a motorway in 3 hours.
   Calculate the average speed. Give your answer to 1 decimal place.

3 Anne walks at an average speed of 3.5 km/h for 2 hours 30 minutes.
   How far does she walk?

4 How long will it take a boat sailing at 12 km/h to travel 64 km?

5 The density of a rock is 9.3 g/cm³. Its volume is 60 cm³.
   What is its mass?

6 Calculate the density of a piece of metal with mass 300 g and volume 84 cm³.
   Give your answer to a suitable degree of accuracy.

7 A man walks 10 km in 2 hours 15 minutes.
   What is his average speed in km/h?
   Give your answer to a suitable degree of accuracy.

8 Calculate the mass of a stone of volume 46 cm³ and density 7.6 g/cm³.

9 Copper has a density of 8.9 g/cm³. Calculate the volume of a block of copper of mass 38 g.
   Give your answer to a suitable degree of accuracy.

10 What is the density of gas if a mass of 32 kg occupies a volume of 25 m³?
    Give your answer to a suitable degree of accuracy.

11 A small town in America has a population of 235 and covers an area of 35 km².
    Find the population density (number of people per square kilometre) of the town.

12 A coach left the coach station at 0910 and travelled 72 km in 90 minutes, arriving at Carter castle.

    (a) Calculate its average speed.

    The coach stopped at the castle for $3\frac{1}{2}$ hours and then travelled back at an average speed of 55 km/h.

    (b) What time did it arrive back at the coach station?
         Give your answer to the nearest minute

17 → PLANNING AND COLLECTING

EXERCISE 17.1H

1 State whether the following are primary data or secondary data.
   (a) Weighing packets of sweets
   (b) Using bus timetables
   (c) Looking up holiday prices on the internet
   (d) A GP entering data for a new patient on his records after seeing the patient

2 Lisa is doing a survey and has written this question.

> What colour is your hair?
>
> Black ☐   Brown ☐   Blonde ☐

   (a) Give a reason why this question is unsuitable.
   (b) Write a better version.

3 Steve is doing a survey about his local sports facilities.
   Here is one of his questions.

> How much do you enjoy doing sport?
>
>   1   2   3   4   5

   (a) Give a reason why this question is unsuitable.
   (b) Write a better version.

4 Mia is doing a survey about school lunches. She gives out questionnaires to the first 30 people in the queue for lunch.
   (a) Why is this likely to give a biased sample?
   (b) Describe a better method of obtaining a sample for her survey.

5 Here is one of Mia's questions.

> Don't you agree that we don't have enough salads on the menu?

   (a) Give a reason why this question is unsuitable.
   (b) Write a better version.

6 A survey is to be done about school students' earnings and pocket money.
   Write five suitable questions which should be included in such a survey.

18 → SEQUENCES

EXERCISE 18.1H

1 Look at this sequence of circles.
The first four patterns in the sequence have been drawn.

  (a) How many circles are there in the 100th pattern?

  (b) Describe the position to term rule for this sequence.

2 Look at this sequence of matchstick patterns.

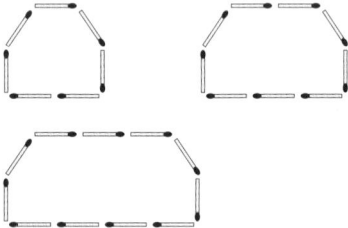

  (a) Complete this table.

Pattern number	1	2	3	4	5
Number of matchsticks					

  (b) What patterns have you noticed in the numbers?

  (c) Find the number of matchsticks in the 50th pattern.

3 Here is a sequence of star patterns.

  (a) Draw the next pattern in the sequence.

  (b) Without drawing the pattern, find the number of stars in the 8th pattern.
Explain how you found your answer.

4 The numbers in a sequence are given by this rule:

   Multiply the position number by 7, then subtract 10.

  (a) Show that the first term of the sequence is -3.

  (b) Find the next four terms in the sequence.

5 Find the first four terms of the sequences with these nth terms.

  (a) $10n$ (b) $8n + 2$

6 Find the first five terms of the sequences with these nth terms.

  (a) n^2 (b) $2n^2$ (c) $5n^2$

7 The first term of a sequence is 3.
The general rule for the sequence is multiply a term by 3 to get to the next term.
Write down the first five terms of the sequence.

8 For a sequence, $T_1 = 12$ and $T_{n+1} = T_n - 5$.
Write down the first four terms of this sequence.

9 Draw suitable patterns to represent this sequence.

   1, 4, 7, 10, ...

10 Draw suitable patterns to represent this sequence.

   $1 \times 1, 3 \times 3, 5 \times 5, 7 \times 7, ...$

EXERCISE 18.2H (pt)

1 Find the nth term for each of these sequences.
  (a) 10, 13, 16, 19, 22, ...
  (b) 0, 1, 2, 3, 4, ...
  (c) −3, −1, 1, 3, 5, ...

2 Find the nth term for each of these sequences.
  (a) 25, 20, 15, 10, 5, ...
  (b) 4, 2, 0, −2, −4, ...
  (c) 3, 2, 1, 0, −1, ...

3 Which of these sequences are linear?
  Find the next two terms of each of the sequences that are linear.
  (a) 2, 5, 10, 17, ...
  (b) 2, 5, 8, 11, ...
  (c) 1, 3, 6, 10, ...
  (d) 12, 8, 4, 0, −4, ...

4 **(a)** Write the first five terms of the sequence with nth term $100n$.
  (b) Compare your answers with this sequence.

    99, 199, 299, 399, ...

  Write down the nth term of this sequence.

5 A mail-order shirt company charges £25 per shirt, plus an overall delivery charge of £3.
  (a) Complete the table.

Number of shirts	1	2	3
Cost in £			

  (b) Write an expression for the cost, in pounds, of n shirts.
  (c) Paul pays £128 for shirts. How many does he buy?

6 **(a)** Write down the first five terms of the sequence with nth term n^2.
  (b) Compare your answers with this sequence.

    0, 3, 8, 15, 24, ...

  Write the nth term of this sequence.

7 The nth triangle number is $\dfrac{n(n+1)}{2}$.

  Find the 60th triangle number.

8 The nth term of a sequence is 2^n.
  (a) Write down the first five terms of this sequence.
  (b) Describe the sequence.

9 **(a)** Write down the first five cube numbers.
  (b) Compare this sequence with the cube numbers.

    3, 10, 29, 66, 127, ...

  Use what you notice to write down the nth term of this sequence.
  (c) Find the 10th term of this sequence.

10 **(a)** Compare this sequence with square numbers.

    5, 20, 45, 80, 125, ...

  Use what you notice to write down the nth term of this sequence.
  (b) Find the 20th term of this sequence.

EXERCISE 19.1H

1 Two points, A and B, are 7 cm apart.
Construct the locus of points that are equidistant from A and B.

2 A badger will never go further than 3 miles from its home.
Construct a scale diagram to show the regions where the badger might go looking for food.

3 Draw an equilateral triangle, ABC, of side 6 cm.
Shade the region of points inside the triangle which are nearer to AB than to AC.

4 A rectangular garden measures 8 m by 6 m.
A fence is built from F, at a right angle across the garden.
Draw a scale diagram and construct the line of the fence.

8 m
6 m
←—5 m—→ F

5 Draw a square, ABCD, of side 5 cm.
Construct the locus of points, inside the square, which are more than 3 cm from A.

6 Zeke is walking across a field.
He notices a bull starting to chase after him.
He runs the shortest distance to the hedge.
Copy the diagram and construct the path that Zeke should run.

Hedge

•Zeke

7 Draw a rectangle, PQRS, with sides PQ = 7 cm and QR = 5 cm.
Shade the region of points that are closer to P than to Q.

8 Draw an angle of 80°.
Construct the bisector of the angle.

9 An office is a rectangle measuring 16 m by 12 m. There are two electricity points in the office at opposite corners of the room. The vacuum cleaner has a wire 10 m long.
Make a scale drawing to show how much of the room can be cleaned.

10 Make another scale drawing of the office in question **9**.
Shade the locus of points which are equidistant from the two electricity points.
Use this locus to work out the length of wire needed for the vacuum cleaner to reach everywhere in the office.

EXERCISE 19.2H

1 Draw a point and label it P.
   Construct the locus of points that are less than
   4 cm from P and more than 6 cm from P.

2 A rectangular garden measures 20 m by 12 m.
   A tree is to be planted so that it is more than 4 m
   from each corner of the garden.
   Make a scale drawing to find the area where
   the tree can be planted.

3 Two points, A and B, are 5 cm apart.
   Find the region that is less than 3 cm from A
   and more than 4 cm from B.

4 Make an accurate drawing of a triangle, PQR,
   where PQ = 6 cm, P = 40° and Q = 35°.
   Find the point X, which is 2 cm from R and
   equidistant from P and Q.

5 Two coastguard stations, A and B, are 20 km
   apart on a straight coastline.
   The coastguard at A knows that a ship is
   within 15 km of him.
   The coastguard at B knows that the same ship
   is within 10 km of him.
   Make a scale drawing to show the region
   where the ship could be.

   A ←— 20 km —→ B

6 Two lines, each 6 cm long, join to form a right
   angle.
   Draw the region of points which are less than
   3 cm from these lines.

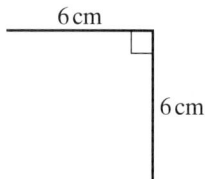

   6 cm
   ⌐
   6 cm

7 Two points, P and Q, are 7 cm apart.
   Find the points which are the same distance
   from P and Q and are also within 5 cm of Q.

8 The diagram shows three coastguard stations,
   C, D and E.
   A ship is within 25 km of C and closer to DE
   than DC.
   Find the region where the ship could be.

   C
   ↑
   20 km
   ↓
   D ←— 25 km —→ E

9 A garden is a rectangle, ABCD, with
   AB = 5 m and BC = 3 m.
   A new flower bed is to be made in the garden.
   It must be more than 2 m from A and less than
   1.5 m from CD.
   Make a scale drawing to show where the
   flower bed could be.

10 EFG is a triangle with EF = 6 cm, FG = 8 cm
   and EG = 10 cm.
   Construct the perpendicular from F to EG.
   Indicate the points on this line that are more
   than 7 cm from G.

EXERCISE 20.1H 🖩 pt

Work these out on your calculator without writing down the answers to the middle stages.

If the answers are not exact, give them correct to 2 decimal places.

1 $\dfrac{7.3 + 8.5}{5.7}$

2 $\dfrac{158 + 1027}{125}$

3 $\dfrac{6.7 + 19.5}{12.2 - 5.7}$

4 $\sqrt{128 - 34.6}$

5 $5.7 + \dfrac{1.89}{0.9}$

6 $(12.6 - 9.8)^2$

7 $\dfrac{8.9}{2.3 \times 5.6}$

8 $\dfrac{15.4}{2.3^2}$

9 $10.9 \times (7.2 - 5.8)$

10 $\dfrac{4.8 + 6.2}{5.2 \times 6.5}$

11 $\dfrac{7.1}{\sqrt{15.3 \times 0.6}}$

12 $\dfrac{3 - \sqrt{2.73 + 5.1}}{4}$

EXERCISE 20.2H

Do not use your calculator for questions **1** to **4**.

1 These calculations are all wrong. This can be spotted quickly without working them out. For each one, give a reason why it is wrong.
   (a) $15.3 \div -5.1 = 5$
   (b) $8.7 \times 1.6 = 5.4375$
   (c) $4.7 \times 300 = 9400$
   (d) $7.5^2 = 46.25$

2 These calculations are all wrong. This can be spotted quickly without working them out. For each one, give a reason why it is wrong.
   (a) $5.400 \div 9 = 60$
   (b) $-6.2 \times -0.5 = -93.1$
   (c) $\sqrt{0.4} = 0.2$
   (d) $8.5 \times 7.1 = 60.36$

3 Estimate the answers to each of these calculations. Show your working.
   (a) 93×108 (b) 0.61^2
   (c) $-19.6 + 5.2$

4 Estimate the answers to these calculations. Show your working.
   (a) The cost of three DVDs at £17.99.
   (b) The cost of 39 cinema tickets at £6.20.
   (c) The cost of five meals at £7.99 and two drinks at £2.10.

🖩 You may use your calculator for questions **5** to **9**.

5 Use inverse operations to check these calculations. Write down the operations you use.
   (a) $19\,669.5 \div 235 = 83.7$
   (b) $\sqrt{5069.44} = 71.2$
   (c) $9.7 \times 12.4 = 120.28$
   (d) $17.2 \times 4.6 + 68.2 = 147.32$

6 Work these out. Round your answers to 2 decimal places.
   (a) $\dfrac{24.3 + 18.6}{2.8 \times 0.51}$
   (b) $(13.7 + 53.1) \times (9.87 - 5.9)$

7 Work these out. Round your answers to 3 decimal places.
   (a) $\dfrac{77.8}{6.4 + 83.9}$
   (b) $1.06^4 \times 185$

8 Work these out. Round your answers to 2 significant figures.
   (a) $\sqrt{2.5^2 + 9.0}$
   (b) 640×0.078

9 (a) Use rounding to 1 significant figure to estimate the answer to each of these calculations. Show your working.

(i) 21.2^3 **(ii)** 189×0.31

(iii) $\sqrt{11.1^2 - 4.8^2}$ **(iv)** $\dfrac{51.8 + 39.2}{0.022}$

(b) Use your calculator to find the correct answer to each of the calculations in **(a)**. Where appropriate, round your answer to a sensible degree of accuracy.

EXERCISE 20.3H

1 Write each of these times as a decimal.
   (a) 8 hours 39 minutes
   (b) 5 hours 21 minutes
   (c) 33 minutes
   (d) 3 minutes

2 Write each of these times in hours and minutes.
   (a) 4.8 hours   **(b)** 5.85 hours
   (c) 0.45 hours   **(d)** 0.6 hours

3 (a) A walker covers a distance of 7.2 miles in 2 hours and 24 minutes.
      Calculate the average speed of the walker in miles per hour.
   (b) For one 25-minute section of its journey an express train travels at an average speed of 126 miles per hour.
      How long is this section of the journey? Give your answer in miles.
   (c) In one stage of its flight a rocket travelled at a constant speed of 357 miles per hour for a distance of 23.8 miles. How long did it take the rocket to travel that distance?

4 In a four-part endurance race, a driver takes the following times to complete each part of the race.

     5 hours 38 minutes
     6 hours 57 minutes
     5 hours 19 minutes
     5 hours 46 minutes

What was the driver's total time for the race? Give your answer in hours and minutes.

5 Wayne buys some potatoes at £1.25 a kilogram and six nectarines at 37p each. He gives the shop assistant £10 and gets £4.68 change. What weight of potatoes did he buy?

6 Chelsea, Sally and James share the profits from their business in the ratio 4 : 3 : 2. In 2005 the total profit was £94 500. Calculate how much Sally received.

7 Merry followed a recipe for lemon pudding which used 350 g of flour for four people. He made the recipe for 10 people and used a new 1.5 kg bag of flour. How much flour did he have left?

8 Freeville has a population of 36 281 and its area is 27.4 km². Calculate its population density. Give your answer to a sensible degree of accuracy.

9 Mr Brown's mobile phone bill one month showed that he had used 53 minutes of calls at 13p per minute. His monthly rental charge for the phone was £15.30. There was VAT at 17.5% on the whole bill. Calculate the total bill including VAT.

10 In January 2000, the Retail Price Index (RPI) was 166.6. In January 2001 it was 171.1. Calculate the percentage increase in the RPI over that year.

11 In September 2003, the Average Earnings Index (AEI) was 113.9. During the next year it increased by 4.2%. Calculate the AEI in September 2004.

12 A metal cylinder has radius 3.8 cm and height 5.7 cm. Its density is 12 g/cm³. Calculate its mass.

13 A water urn is in the shape of a cuboid, with the base a square of side 30 cm. How many litres of water does the urn contain when it is filled to a depth of 42 cm?

14 David drove 28 miles along the motorway at 70 mph, and then 10 miles at 50 mph.
   (a) Calculate how long he took.
   (b) Find his average speed for the whole journey.

EXERCISE 21.1H

1   Find the gradient of each of these lines.

(a)

(b)

(c)

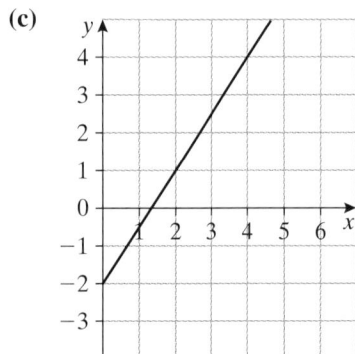

2   Find the gradient of the line joining each of these pairs of points.
   (a) (1, 2) and (3, 8)
   (b) (5, 1) and (7, 3)
   (c) (0, 3) and (2, −3)
   (d) (−1, 4) and (3, 2)
   (e) (3, −1) and (−1, −1)

3   Find the gradient of each side of the triangle ABC.

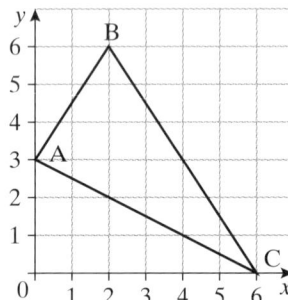

4   Find the gradient of each of these lines.

(a)

(b)

(c)

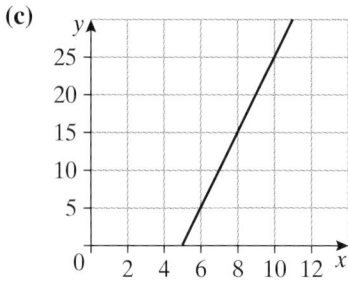

5 Draw the graph of each of these straight lines and find its gradient.

    (a) $y = 3x + 1$ **(b)** $y = x - 2$

    (c) $y = -3x + 2$ **(d)** $y = -2x - 1$

    (e) $3x + 4y = 12$

6 Find the velocity for each of these distance–time graphs.

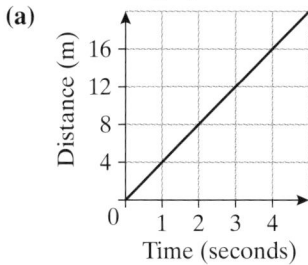

    (a)

    (b)

EXERCISE 21.2H

1 Write down the equations of straight lines with these gradients and y-intercepts.

    (a) Gradient 2, y-intercept 6

    (b) Gradient -3, y-intercept 5

    (c) Gradient 1, y-intercept 0

2 Find the equation of each of these lines.

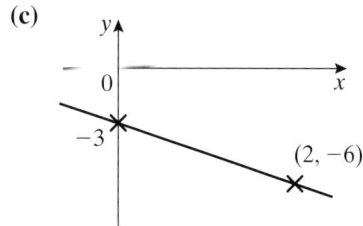

    (a)

    (b)

    (c)

3 Find the gradient and y-intercept of each of these lines.

    (a) $y = 4x - 2$ **(b)** $y = 3x + 5$

    (c) $y = -2x + 1$ **(d)** $y = -x + 2$

    (e) $y = -3.5x - 8$

4 Find the gradient and y-intercept of each of these lines.

    (a) $y + 3x = 7$ **(b)** $6x + 2y = 5$

    (c) $2x + y = 3$ **(d)** $5x - 2y = 8$

    (e) $6x + 4y = 9$

5 Find the equation of each of these lines.

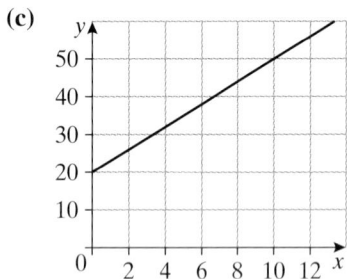

(a)

(b)

(c)

EXERCISE 21.3H

Solve graphically each of these pairs of simultaneous equations.

1 $y = 3x$ and $y = 4x - 2$.
Use values of x from -1 to 4.

2 $y = 2x + 3$ and $y = 4x + 1$.
Use values of x from -2 to 3.

3 $y = x + 4$ and $4x + 3y = 12$.
Use values of x from -3 to 3.

4 $y = 2x + 8$ and $y = -2x$.
Use values of x from -5 to 1.

5 $2y = 3x + 6$ and $3x + 2y = 12$.
Use values of x from 0 to 4.

EXERCISE 21.4H

Use algebra to solve each of these pairs of simultaneous equations.

1 $x + y = 3$
$2x + y = 4$

2 $2x + y = 6$
$2x - y = 2$

3 $2x - y = 7$
$3x + y = 13$

4 $2x + y = 12$
$x - y = 3$

5 $2x + y = 7$
$3x - y = 8$

6 $x + y = 4$
$3x - y = 8$

7 $x + 3y = 9$
$2x - 3y = 0$

8 $3x + y = 14$
$3x + 2y = 22$

9 $4x - y = 4$
$4x + 3y = 20$

10 $x - 4y = 2$
$x + 3y = 9$

EXERCISE 21.5H pt

Use algebra to solve each of these pairs of simultaneous equations.

1 $x + 3y = 5$
$2x + y = 5$

2 $2x - 5y = 3$
$x + y = 5$

3 $3x - y = 3$
$2x + 3y = 13$

4 $4x - y = 2$
$5x + 3y = 11$

5 $3x - 2y = 8$
$2x - y = 5$

6 $2x + y = 5$
$7x + 2y = 13$

7 $x + 3y = 4$
$3x + 2y = -2$

8 $4x + 3y = 11$
$x + 2y = 4$

9 $x + 2y = 8$
$2x - 3y = 9$

10 $x + y = 2$
$x + 3y = 5$

EXERCISE 22.1H

1 Calculate how much these items are worth if they increase by the given fraction each year for the given number of years.
Give your answers to the nearest penny.

	Original value	Fractional increase	Number of years
(a)	£5000	$\frac{1}{30}$	4
(b)	£300	$\frac{2}{7}$	3
(c)	£4500	$\frac{5}{9}$	6

2 Calculate how much these items are worth if they decrease by the given fraction each year for the given number of years.
Give your answers to the nearest penny.

	Original value	Fractional decrease	Number of years
(a)	£120	$\frac{1}{6}$	5
(b)	£5200	$\frac{3}{5}$	6
(c)	£140	$\frac{1}{2}$	4

3 Cathy invested £1870 with a bond that offered to increase the amount by $\frac{1}{12}$ each year.
How much was the bond worth after 4 years?

4 At Patnik shoe shop they offered to reduce the price of a pair of shoes by $\frac{1}{4}$ each day until they were sold.
They were priced at £47 to start with.
What was the price after 4 reductions?
Give your answer to the nearest penny.

5 It is estimated that the royalties from a book will decrease by $\frac{2}{5}$ each year.
In 2004 Kath received £11 000 in royalties. If the estimate is correct, how much will she receive 5 years later?

EXERCISE 22.2H pt

1 Copy and complete this table.

	Original value	Percentage increase	Increased value
(a)	£750	8%	
(b)		15%	£414
(c)	£42.50	4.5%	
(d)		5%	£254

2 Copy and complete this table.

	Original value	Percentage decrease	Decreased value
(a)	£2000	12%	
(b)		5%	£240
(c)	£260	3.5%	
(d)		12.5%	£325

3 After an increase of 12%, a quantity is 84 tonnes.
What was it before the increase?

4 A newspaper increased its circulation by 3% and the new number sold was 58 195.
What was the circulation before the increase?

5 Santos bought a car for £14 750.
He sold it three years later at a loss of 45%.
What price did he sell it for?

6 A charity's income has been reduced by 2.5%.
Its income is now £8580.
What was it before the reduction?

7 It was announced that the number of people
unemployed had decreased by 3%.
The number unemployed before the decrease
was 2.56 million.
How many are now unemployed?

8 The price of a car is £12 925, including VAT
at 17.5%.
What is the price without VAT?

9 At Percival's sale the price of everything was
reduced by 7.5%, rounded to the nearest
penny.
(a) A pair of boots cost £94.99 before the sale.
What was the price in the sale?
(b) Delia was charged £13.87 for a blouse in
the sale.
What was its original price?

10 John's pension has increased by 4.75% and is
now £924.56 a month.
What was it before the increase?

11 At St Bede's church they give $\frac{1}{8}$ of their
weekly collection to overseas aid.
One week, they had £151.20 left after they had
given away the money.
How much was the total collection?

12 The makers of Steamer's jam say that the new
size jar contains a fifth more jam than the old
size jar.
The new size jar contains 570 g.
How much did the old size jar contain?

EXERCISE 23.1H (pt)

1 A call centre records the calls its operators make. The table shows the duration of the calls recorded one day.

Duration (minutes)	Number of calls
0 to 1	230
1 to 2	420
2 to 3	480
3 and over	358

50 calls are to be selected at random, to be reviewed by the centre manager.
How many calls should be selected from each band to make the sample representative?

2 A simple random sample of 20 households in a particular street of a town are to be surveyed about their refuse collection. The houses in the street are numbered 1 to 60.

Use the random number table below to choose the sample. Start at the top left corner and work your way along the first and subsequent rows, examining pairs of digits in the table. Write down the first 20 pairs which lie between 01 and 60, ignoring repetitions and those values outside the range.

980677	461663	998081	821548	961256
402566	215166	163433	183641	331870
685871	249206	948448	929632	290060
783289	766103	012094	363987	522723

3 A systematic random sample of 20 households in the same street as in question **2** are to be surveyed about how often they travel by bus.

Since 20 out of 60 is the same as 1 in 3, we need to randomly select a starting value between 1 and 3. Starting at the beginning of row 2 of the random number table above, look at single digits and find the first value which lies between 1 and 3. Use this as the start of your systematic sample.

Write down the 20 house numbers that this method gives.

4 The students of a school are to be surveyed about a new school uniform.
100 students are to be interviewed.
The table shows the number of students in each year group.

Year	7	8	9	10	11
Number of students	142	154	115	127	102

How many students from each year group should be selected to provide a representative sample?

EXERCISE 23.2H

In each of these questions, decide whether or not
the method of sampling is appropriate. If it is not
satisfactory, say why not.

1   A candidate in a local election wishes to
    establish how people are likely to vote. He
    decides to telephone every 50th person in the
    telephone directory.

2   To obtain information about diseases amongst
    the elderly, everyone in a large residential
    home is given a thorough health check.

3   The Post Office is interested to know how
    people feel about proposed changes to
    delivery times for letters. They send a letter to
    every household and ask their views.

4   A manufacturer is about to introduce a new
    flavour of crisps. They stop people in the street
    and offer them some of the crisps to taste.

5   To find people's views about an increase in
    income tax, a random sample of ten people are
    stopped and questioned in the street one
    morning.

EXERCISE 24.1H pł

1   Find the length of each of the lines in the diagram.
Where the answer is not exact, give your answer correct to 2 decimal places.

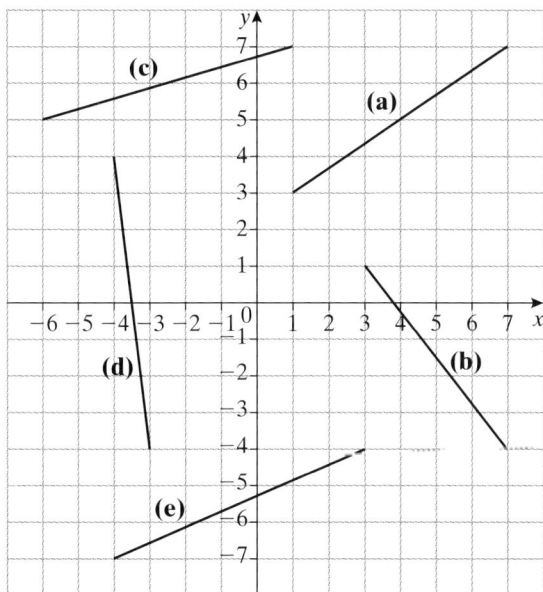

2   Find the length of the line joining each of these pairs of points.
You can draw a diagram to help you.
Where the answer is not exact, give your answer correct to 2 decimal places.
(**a**) A(2, 7) and B(6, 4)   (**b**) C(5, 8) and D(7, 2)
(**c**) E(3, 8) and F(7, −1)   (**d**) G(6, 5) and H(−6, 0)
(**e**) I(−3, 2) and J(−8, −5)   (**f**) K(−7, 2) and L(3, 6)

EXERCISE 24.2H pt

1   In these diagrams find the lengths marked a, b, c, d, e, f, g and h.

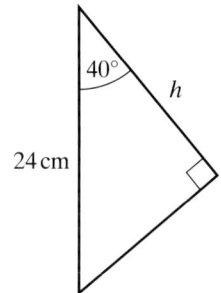

2   The diagram shows the side view of a waste bin.
   (a) Find the width, w cm, of the bin.
   (b) Find the height, h cm, of the bin.

3   **(a)** Find the height, a cm, of the triangle.
   (b) Find the length, b cm.
   (c) Find the length, c cm.
   (d) Use your answers to parts **(a)**, **(b)** and **(c)** to find the
         area of the triangle.

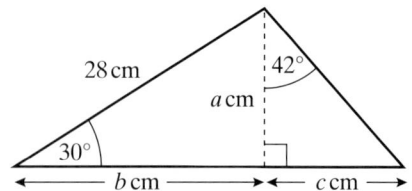

EXERCISE 24.3H pt

1   In these diagrams find the lengths marked *a*, *b*, *c*, *d*, *e*, *f*, *g* and *h*.

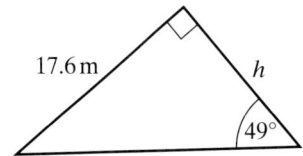

2   The diagram shows a step ladder.
The two sections of the ladder are opened to 30°.
The feet of the two parts are 1.2 m apart.
Calculate the length, *l*, of each section of the ladder.

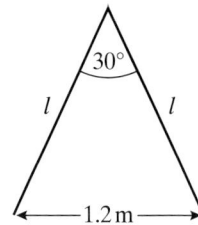

3   **(a)** Find the base, *a* cm, of the triangle.
   (b) Use the base you found in part **(a)** to find the area
      of the triangle.

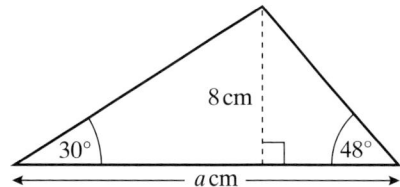

EXERCISE 24.4H

1 In these diagrams find the angles marked *a*, *b*, *c*, *d*, *e*, *f*, *g* and *h*.

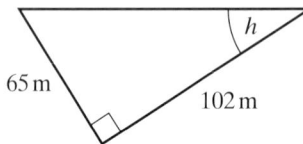

16 cm, *a*, 11 cm

16 cm, 19 cm, *b*

0.8 m, 5.3 m, *c*

11.6 cm, 13.2 cm, *d*

e, 8.3 cm, 12.1 cm

6.1 m, 7.6 m, *f*

6.8 cm, *g*, 21.8 cm

65 m, *h*, 102 m

2 A rectangle has a length of 12 cm and a width of 8 cm.
Calculate the angle the diagonal makes with the longest side.

3 A yacht sails from A to B.
B is 43 km east of A and 25 km south of A.
Calculate the bearing of B from A.

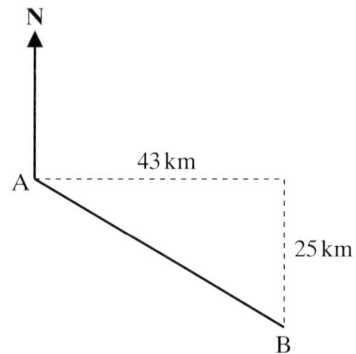

N

A, 43 km, 25 km, B

4 The diagram shows an isosceles triangle.
Calculate the angle *x*.

x, 8 cm, 8 cm, 13 cm

25 ➔ REPRESENTING AND INTERPRETING DATA

EXERCISE 25.1H (pt)

1 This cumulative frequency
graph shows the heights
of 80 sunflowers.
Find the median, the
quartiles and the
interquartile range of
these heights.

2 The table on the left shows information about the masses of 140 tomatoes.
   (a) Copy and complete the cumulative frequency table on the right.

Mass (m grams)	Frequency
$0 < m \leqslant 20$	4
$20 < m \leqslant 40$	14
$40 < m \leqslant 60$	25
$60 < m \leqslant 80$	47
$80 < m \leqslant 100$	36
$100 < m \leqslant 120$	14

Mass (m grams)	Cumulative frequency
$m \leqslant 0$	0
$m \leqslant 20$	4
$m \leqslant 40$	18
$m \leqslant 60$	
$m \leqslant 80$	
$m \leqslant 100$	
$m \leqslant 120$	

   (b) Draw the cumulative frequency graph.
   (c) Use your graph to find the median and interquartile range of these masses.

3 The table on the left shows the ages of people in a health club.

Age (years)	Frequency
under 11	32
11–18	25
19–29	53
30–49	83
50–69	45
70–95	21

Age (years)	Cumulative frequency
$y < 0$	0
$y < 11$	32
$y < 19$	57
$y < 30$	110
$y <$	
$y <$	

(a) Copy and complete the cumulative frequency table on the right.
   Note: the upper boundary of the 11–18 age group is the 19th birthday.
(b) Draw the cumulative frequency graph.
(c) How many people in this club are aged under 40?
(d) How many people in this club are aged 60 or over?
(e) Find the median and the quartiles.

4 The cumulative frequency graph shows the foot lengths of a sample of 50 boys and 50 girls.

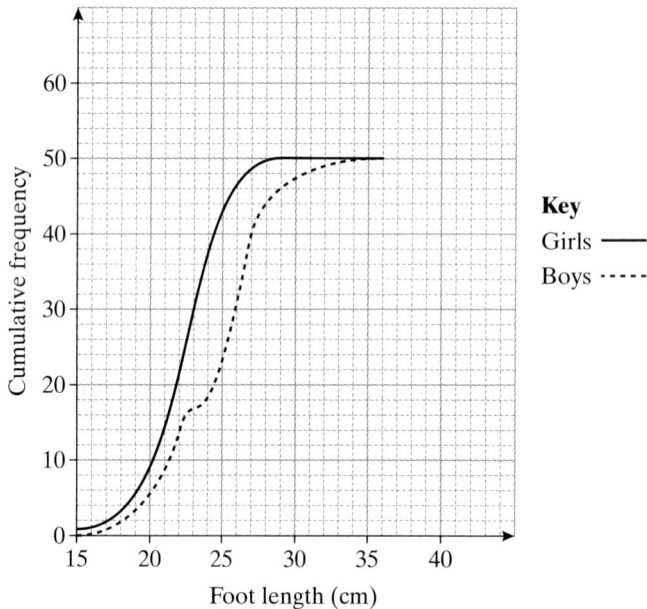

Key
Girls ——
Boys ·····

(a) What does the flat section at the top of the girls' graph tell you?
(b) Compare the distributions. Make two comparisons.

5 The stem-and-leaf diagram represents the marks gained by 35 students in a test.

```
3 | 2 5
4 | 2 3 6 8
5 | 0 1 2 3 7
6 | 1 2 4 4 7 8 8
7 | 0 0 1 2 3 4 5 6 7 9
8 | 1 3 4 6 7
9 | 3 5
```

Key: 4 | 7 represents 47 marks.

Copy and complete this table for the distribution.

Minimum (Q_0)	LQ (Q_1)	Median (Q_2)	UQ (Q_3)	Maximum (Q_4)

6 This table shows the distance walked by each of a group of students one day.

Distance (d miles)	Frequency
$0 < d \leqslant 2$	4
$2 < d \leqslant 4$	16
$4 < d \leqslant 6$	8
$6 < d \leqslant 8$	6
$8 < d \leqslant 10$	2

Draw a cumulative frequency graph and a box plot to represent this distribution.

7 Here is a summary for a distribution of marks gained by a year-group in an examination.

Minimum (Q_0)	LQ (Q_1)	Median (Q_2)	UQ (Q_3)	Maximum (Q_4)
27	45	62	74	93

Draw a box plot for this distribution.

8 The cumulative frequency graph represents the distances swum by children in a sponsored swim.

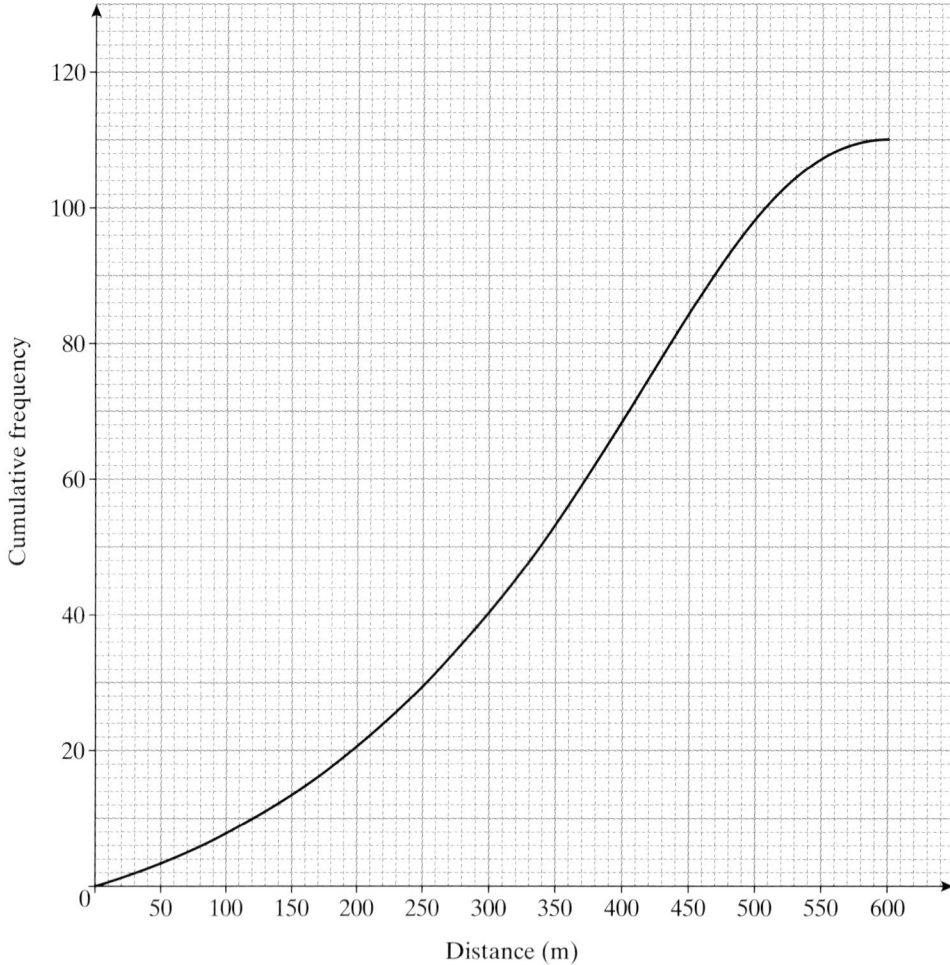

Distance (m)

(a) How many children took part in the swim?
(b) How many children swam over 400 m?
(c) Draw a box plot for this distribution.

9 These box plots represent the time, in hours, spent watching TV by samples of 50 boys and 50 girls one week.
Make two comparisons between the times spent by the boys and the girls, stating which statistics you use.

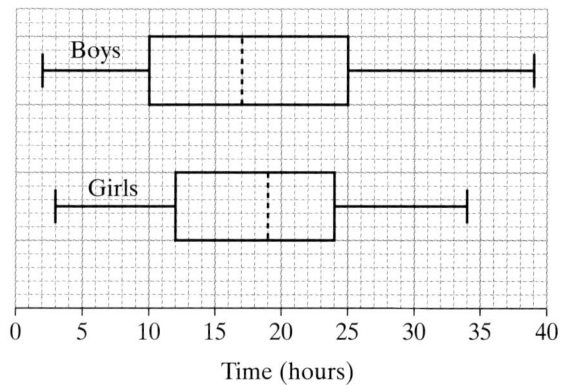

Time (hours)

10 A company tested a sample of 200 torch batteries of each of two types it produces.
The table summarises the results, showing the time, in hours, that each battery lasted.

Time (t hours)	Frequency for type A	Frequency for type B
$0 < t \leq 5$	4	12
$5 < t \leq 10$	31	61
$10 < t \leq 15$	45	71
$15 < t \leq 20$	87	32
$20 < t \leq 25$	27	16
$25 < t \leq 30$	6	8

  (a) On the same axes, draw cumulative frequency graphs to represent these distributions.
Below your graphs, draw box plots for the distributions.

  (b) Which of the two types of battery is more reliable?

EXERCISE 25.2H (pt)

1 The table shows the amounts spent at the supermarket by a sample of people.

Amount spent (£s)	$0 < s \leq 20$	$20 < s \leq 40$	$40 < s \leq 70$	$70 < s \leq 100$	$100 < s \leq 150$
Frequency	12	16	33	12	6

Draw a histogram to represent this distribution. Label your vertical scale or key clearly.

2 This distribution shows the ages of people visiting a swimming pool one day.

Age (years)	Under 10	10–19	20–29	30–49	50–89
Frequency	96	58	36	58	144

  (a) Explain why the boundary of the 30–49 group is 50 years.

  (b) Calculate the frequency densities and draw a histogram to represent this distribution.

3 The histogram represents a distribution of waiting times for non-emergency operations at a particular hospital.

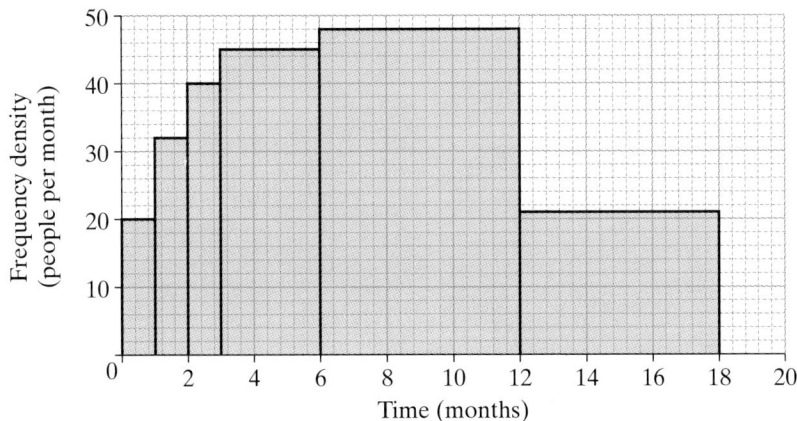

  (a) Make a frequency table for this distribution.

  (b) Calculate an estimate of the mean waiting time.

4 The histograms represent the times spent on a piece of coursework by a sample of girls and boys.

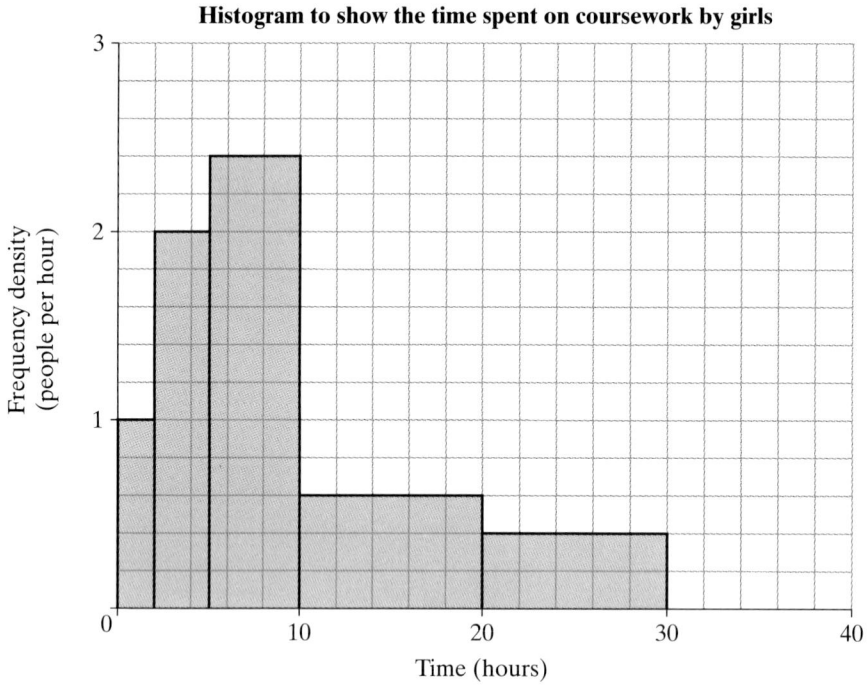

Histogram to show the time spent on coursework by girls

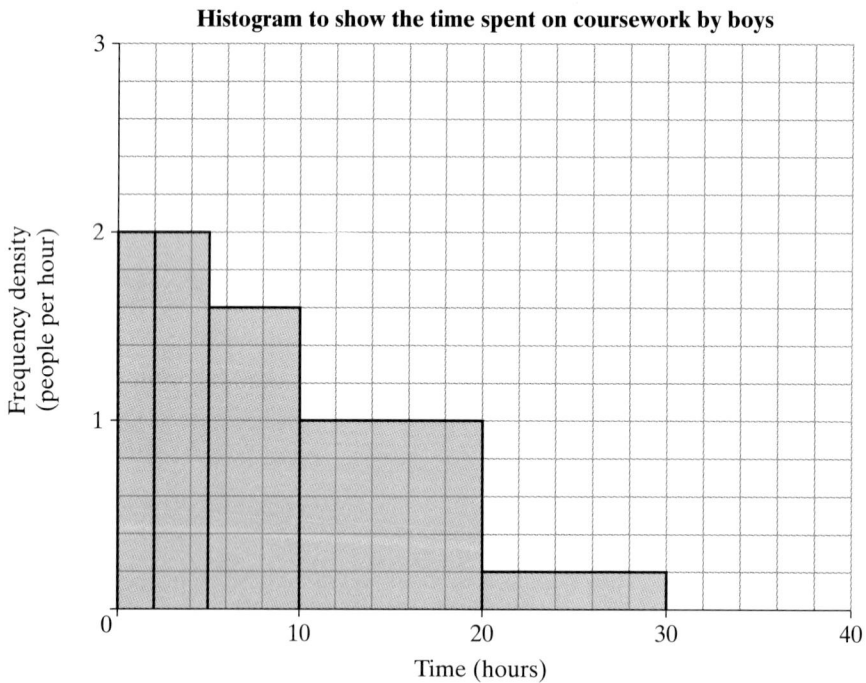

Histogram to show the time spent on coursework by boys

(a) Find how many girls and how many boys spent between 5 and 10 hours on the coursework.

(b) Compare the distributions.

5 The members of a gym were measured. The histogram represents their heights.

(a) How many members of the gym were measured?

(b) Calculate an estimate of their mean height.

EXERCISE 25.3H

1 The table shows the number of people visiting a cinema complex each day for a 4-week period.

	Su	M	Tu	W	Th	F	Sa
Week 1	2135	1051	1604	2015	1752	1854	3045
Week 2	2614	1252	1527	1927	1813	1927	2984
Week 3	2542	1306	1614	1852	1782	2016	3257
Week 4	2968	1422	1827	2094	1905	2283	3371

(a) Plot a time-series graph of these figures.

(b) Calculate the 7-day moving averages and add these to the graph.

(c) Add a trendline. Describe the trend.

2 The table shows the number of letters delivered by a postman each day for a 4-week period.

	M	Tu	W	Th	F	Sa
Week 1	201	205	304	196	316	92
Week 2	192	216	296	142	301	115
Week 3	185	205	310	185	298	106
Week 4	170	186	284	135	277	72

(a) Plot a time-series graph of these figures.

(b) Calculate the 6-day moving averages and add these to the graph.

(c) Describe the trend over this period.

3 These figures show the number of mobile text messages sent by the Peters' family over a period of 3 years.

	1st quarter	2nd quarter	3rd quarter	4th quarter
Year 1	210	192	361	242
Year 2	262	352	572	391
Year 3	354	529	834	682

(a) Describe the seasonal variation.
(b) Calculate the 4-point moving averages.
(c) Describe the trend.

4 The graph shows a shop's quarterly sales of ice-cream.

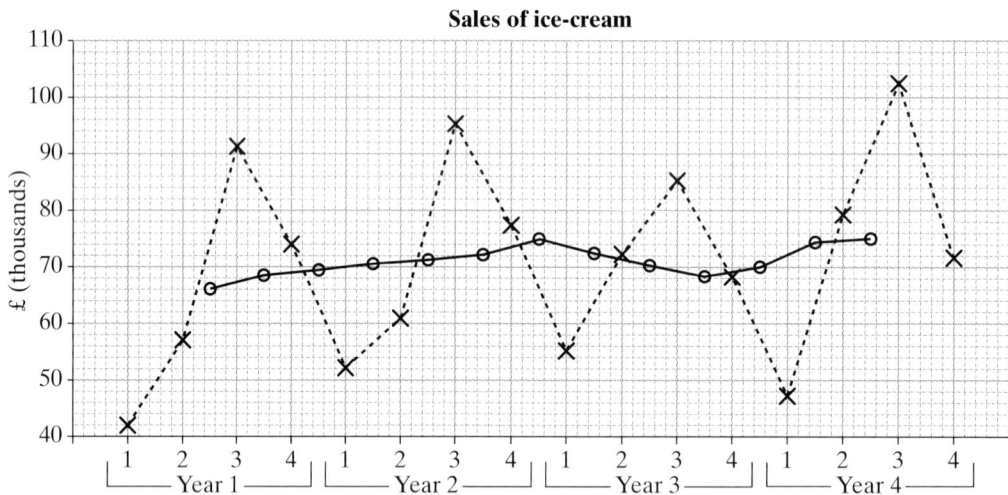

Sales of ice-cream

(a) Use readings from this graph to show how the third moving average has been calculated.
(b) Comment on the seasonal variation shown on the graph.
(c) Which summer had disappointing sales?

26 → INDICES AND STANDARD FORM

✗ EXERCISE 26.1H pt

1 Write these in index form.
   (a) $\sqrt[5]{x}$
   (b) The reciprocal of $x^{\frac{1}{2}}$
   (c) $\sqrt[4]{x^3}$

2 Work out these. Give your answers as whole numbers or fractions.
   (a) 25^{-1} (b) $25^{\frac{1}{2}}$ (c) 25^0
   (d) $25^{-\frac{1}{2}}$ (e) $25^{\frac{3}{2}}$ (f) $27^{\frac{2}{3}}$
   (g) $10\,000^{\frac{1}{4}}$ (h) $\left(\frac{1}{100}\right)^{-\frac{1}{2}}$ (i) $32^{\frac{6}{5}}$
   (j) $\left(\frac{1}{2}\right)^0$ (k) $16^{\frac{1}{2}}$ (l) 16^0
   (m) $16^{\frac{3}{2}}$ (n) $16^{-\frac{1}{4}}$ (o) $16^{\frac{7}{4}}$
   (p) $144^{\frac{1}{2}}$ (q) $\left(\frac{2}{5}\right)^{-2}$

3 Work out these. Give your answers as whole numbers or fractions.
   (a) $1000^{\frac{2}{3}} \times 8^{\frac{2}{3}}$
   (b) $100^{-\frac{1}{2}} \times 49^{\frac{3}{2}}$
   (c) $4^{-2} \times 10^4 \times 25^{-\frac{1}{2}}$
   (d) $4^3 + 16^{\frac{1}{2}} - \left(\frac{1}{5}\right)^{-2}$
   (e) $10\,000^{\frac{1}{4}} + 125^{\frac{1}{3}} - 121^{\frac{1}{2}}$
   (f) $\left(\frac{4}{5}\right)^2 \times 128^{-\frac{3}{7}}$

▦ EXERCISE 26.2H pt

1 Work out these. Give your answers exactly or to 5 significant figures.
   (a) 5.3^4 (b) 0.72^5
   (c) 1.03^7 (d) 1.37^{-3}
   (e) $28\,561^{\frac{1}{4}}$ (f) $9.23^{\frac{1}{5}}$
   (g) $\sqrt[4]{93}$ (h) $2187^{\frac{3}{7}}$

2 Work out these. Give your answers exactly or to 5 significant figures.
   (a) 300×1.05^{10} (b) $2.4^6 \times 1.2^5$
   (c) $5.7^3 \div 2.6^{-3}$ (d) $(4.7 \times 5.1^4)^{\frac{1}{5}}$
   (e) $3.7^4 + 1.3^7$ (f) $3.7^{\frac{1}{4}} - 12.8^{\frac{1}{6}}$
   (g) $2.7^5 + 0.8^{-4}$ (h) $512^{\frac{4}{3}} \div 3125^{\frac{2}{5}}$

✗ EXERCISE 26.3H

1 Write these as powers of 2 as simply as possible.
   (a) 64 (b) $8^{\frac{2}{3}}$ (c) 0.25
   (d) $2 \times \sqrt[3]{64}$ (e) $4^{\frac{n}{2}}$ (f) $2^{3n} \times 4^{\frac{n}{2}}$

2 Where possible, write these as powers of a prime number as simply as possible.
   (a) 343 (b) $9^{\frac{2}{3}}$
   (c) $64^{-\frac{2}{3}}$ (d) $16^{\frac{1}{2}} \times 64^{-\frac{2}{3}}$
   (e) $2^6 + 2^3$ (f) $27 \div 81^{\frac{3}{2}}$
   (g) $9^{3n} \times 3^{-n}$

3 Write each of these in the form $2^a \times 3^b$ or $2^a \times 3^b \times 5^c$.
   (a) 60 (b) 192
   (c) 600 (d) 648

4 Write each of these numbers as a product of powers of prime numbers.
   (a) 15^3 (b) $25^2 \times 10^{\frac{1}{2}}$
   (c) 40^n (d) $40^n \times 10$

EXERCISE 26.4H

1 Write these numbers in standard form.
   (a) 60 000   (b) 8400
   (c) 863 000   (d) 72 500 000
   (e) 9 020 000   (f) 78
   (g) 5.2 million

2 Write these numbers in standard form.
   (a) 0.08   (b) 0.0096
   (c) 0.000 308   (d) 0.000 063
   (e) 0.000 004 8   (f) 0.000 000 023

3 These numbers are in standard form. Write them as ordinary numbers.
   (a) 3×10^3 (b) 4.6×10^4
   (c) 2×10^{-5} (d) 7.2×10^5
   (e) 1.9×10^{-4} (f) 5.78×10^6
   (g) 2.87×10^8 (h) 5.13×10^{-6}
   (i) 2.07×10^{-3} (j) 7.28×10^7

EXERCISE 26.5H pt

1 Work out these. Give your answers in standard form.
   (a) $(6 \times 10^4) \times (3 \times 10^5)$
   (b) $(8 \times 10^6) \div (4 \times 10^3)$
   (c) $(9 \times 10^5) \times (4 \times 10^{-2})$
   (d) $(3 \times 10^{-3}) \times (2 \times 10^{-4})$
   (e) $(2 \times 10^8) \div (5 \times 10^3)$
   (f) $(7.6 \times 10^5) + (3.8 \times 10^4)$
   (g) $(5.6 \times 10^{-3}) - (4 \times 10^{-4})$

2 Work out these. Give your answers in standard form.
   (a) $(4.4 \times 10^6) \times (2.7 \times 10^5)$
   (b) $(6.5 \times 10^6) \times (2.3 \times 10^2)$
   (c) $(7.1 \times 10^4) \times (8.3 \times 10^2)$
   (d) $(2.82 \times 10^4) \div (1.2 \times 10^{-2})$
   (e) $(7.2 \times 10^3) \times (1.3 \times 10^5)$
   (f) $(4.3 \times 10^3) + (6.72 \times 10^4)$
   (g) $(6.21 \times 10^5) - (3.75 \times 10^4)$

27 → SIMILARITY AND ENLARGEMENT

EXERCISE 27.1H (pt)

1 Two rectangles are similar.
The ratio of the corresponding sides of the two rectangles is 2 : 5.
The width of the smaller rectangle is 9 cm and the length of the larger one is 45 cm.
Find
(a) the length of the smaller rectangle.
(b) the width of the larger rectangle.

2 Write down the triangle which is similar to triangle ABC.
Make sure you put the letters in the correct order.

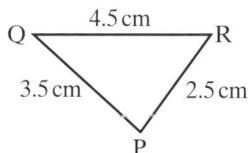

3 For these two triangles, write down the corresponding sides.

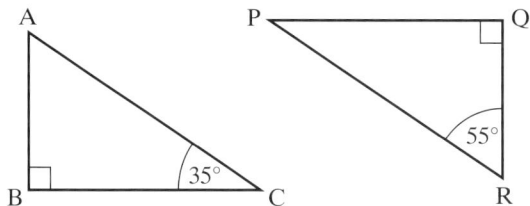

4 Triangle ABC is similar to triangle QRP.

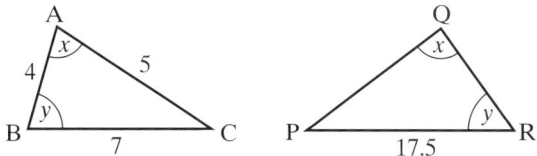

Calculate the lengths of PQ and QR.

5 In triangle ABC, XY is parallel to BC, XA = 4 cm, YA = 5 cm, BX = 2 cm and XY = 2 cm.

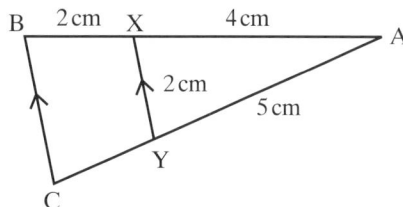

(a) State which triangle is similar to triangle ABC.
(b) Calculate the lengths of BC and CY.

6 Calculate the lengths x and y in this diagram.

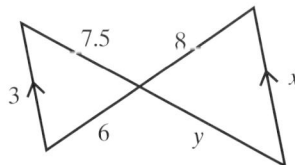

7 Triangles ABC and CBD are similar.
AC = 3 cm, BC = 6 cm and BD = 9 cm.

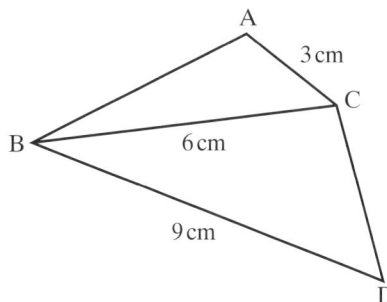

Calculate these lengths.
(a) AB
(b) CD

8 This is a diagram of a triangle with its top cut off.
AB is parallel to XY.

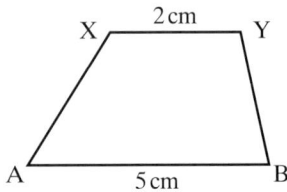

AB = 5 cm and XY = 2 cm.
The height of the piece cut off is 3.5 cm.
Find the height of the complete triangle.

EXERCISE 27.2H

1 Draw a set of axes with the *x*-axis from −8 to 16 and the *y*-axis from −6 to 10.
Plot the points A(−2, −1), B(−2, −4) and C(−7, −1) and join them to form a triangle.
Enlarge the triangle by a scale factor of −2 using the origin as the centre of enlargement.

2 Draw a set of axes with the *x*-axis from −8 to 8 and the *y*-axis from −8 to 8.
Plot the points A(6, 5), B(6, 8) and C(8, 8) and join them to form a triangle.
Enlarge the triangle by a scale factor of −3 using (4, 4) as the centre of enlargement.

3 The diagram shows a quadrilateral, ABCD, and its image A′B′C′D′.

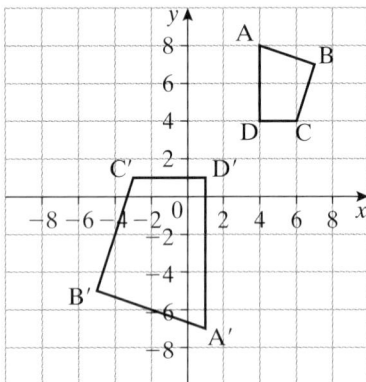

Copy the diagram and find
(a) the centre of enlargement.
(b) the scale factor.

4 Draw a set of axes with the *x*-axis from −10 to 10 and the *y*-axis from −8 to 8.
Plot the points A(−4, 3), B(−4, −3) and C(−10, −5) and join them to form a triangle.
Enlarge the triangle by a scale factor of $-\frac{1}{2}$ using (2, 3) as the centre of enlargement.

5 The diagram shows a triangle, ABC, and its image A′B′C′.

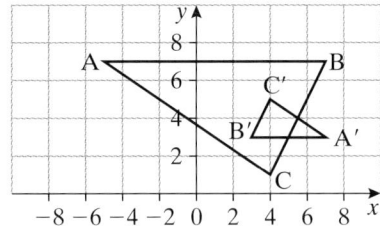

Copy the diagram and find
(a) the centre of enlargement.
(b) the scale factor.

EXERCISE 27.3H

1 A cube has edges 4 cm long. Another cube has edges 12 cm long.
(a) Write down the linear scale factor for the enlargement.
(b) Write down the area of
  (i) a face of the small cube.
  (ii) a face of the big cube.
(c) (i) Write down the area scale factor.
  (ii) What do you notice?
(d) Write down the volume of
  (i) the small cube.
  (ii) the big cube.
(e) (i) Write down the volume scale factor.
  (ii) What do you notice?

2 Triangles ABC and PQR are similar.

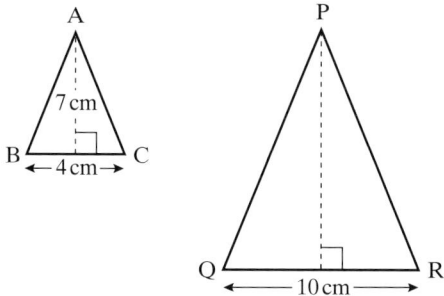

(a) What is the linear scale factor of the enlargement?
(b) Find the height of triangle PQR.
(c) Calculate the area of triangle ABC.
(d) Calculate the area of triangle PQR.
(e) Write down the ratio of the areas.

3 State the area scale factor and the volume scale factor for each of these linear scale factors.
(a) 3   (b) 4   (c) 8
(d) $\frac{1}{2}$ (e) $\frac{5}{3}$

4 State the linear scale factor for each of these area scale factors.
(a) 49   (b) 81   (c) 100
(d) 2500   (e) $\frac{9}{25}$

5 A model boat is built to a scale of 1 : 50.
It is 30 cm long.
(a) How long is the real boat?
(b) The real boat has a mast 8 m high.
How high is the mast on the model?

6 A table is 0.6 m wide and the top has an area of 0.56 m^2.
Another table has a similar shaped top. It is 0.9 m wide.
What is the area of the top of this table?

7 A small bottle holds 150 ml of liquid.
A similar bottle is twice as tall.
How much liquid does it hold?

8 Shirley has a poster made from a photo she has taken.
The poster is an enlargement of the photo with linear scale factor 8.
The dimensions of the photo are 5 cm by 7 cm.
What is the area of the poster?

9 A vase is 12 cm tall.
Another similar vase is 18 cm tall.
The larger vase has a capacity of 54 cm^3.
What is the capacity of the smaller vase?

10 Two cuboids are similar.
The smaller cuboid has edges 4 cm, 5 cm and 8 cm long.
The larger one has a volume of 20 000 cm^3.
What is the length of the shortest edge of the larger cuboid?

28 → ALGEBRAIC MANIPULATION 2

EXERCISE 28.1H pt

Simplify each of these algebraic fractions.

1 $\dfrac{7x - 14}{4x + 8}$

2 $\dfrac{3x - 6}{9 - 12x}$

3 $\dfrac{12 + 8x}{6x - 4}$

4 $\dfrac{18 - 6x}{12 + 9x}$

5 $\dfrac{4x^2 + 2x}{2x^2 - 6x}$

6 $\dfrac{3x^2 - 4x}{5x^2 - 6x}$

7 $\dfrac{8 - 6x}{4x^2 + 6x}$

8 $\dfrac{3x^2 + 6x}{9x - 12x^2}$

EXERCISE 28.2H pt

Factorise each of these expressions.

1 $x^2 + 7x + 6$

2 $x^2 + 8x + 15$

3 $x^2 + 9x + 18$

4 $x^2 + 14x + 13$

5 $x^2 + 10x + 25$

6 $x^2 + 11x + 24$

7 $x^2 + 13x + 40$

8 $x^2 + 19x + 48$

EXERCISE 28.3H

Factorise each of these expressions.

1 $x^2 - 1$

2 $x^2 - 36$

3 $x^2 - 121$

4 $x^2 - 196$

5 $2x^2 - 32$

6 $4x^2 - 100$

7 $6x^2 - 54$

8 $9x^2 - 576$

EXERCISE 28.4H

Simplify each of these expressions.

1 $\dfrac{8a^3b^5}{9c^4} \times \dfrac{15c^2}{4a^5b^3}$

2 $\dfrac{15x^3y^2}{16z} \times \dfrac{12z^3}{25x^4y^5}$

3 $\dfrac{4p^5}{3r^6s^7} \times \dfrac{15r^6s^4}{8p^2}$

4 $\dfrac{7h^4j^2}{6k} \div \dfrac{14h^7j^5}{15k^3}$

5 $\dfrac{20t^6}{9v^5w^2} \div \dfrac{15t^2}{6v^3w^4}$

6 $\dfrac{18a^8b^2}{25c^5} \div \dfrac{27a^5b^4}{20c^9}$

7 $\dfrac{15a^4b^5}{8c^4d^3} \times \dfrac{24b^5c^4}{35a^2d^3} \times \dfrac{7c^3d^2}{6ab^6}$

8 $\dfrac{9w^6x^5}{10yz^2} \div \dfrac{3w^3y^4}{4x^4z^3} \times \dfrac{5w^2z}{6x^5y^6}$

EXERCISE 28.5H (pt)

Factorise each of these expressions.

1 $3x^2 + 17x + 10$

2 $2x^2 + 7x + 3$

3 $2x^2 + x - 21$

4 $5x^2 - 19x - 4$

5 $2x^2 + x - 15$

6 $3x^2 + 11x - 20$

7 $2x^2 - 11x + 12$

8 $5x^2 - 19x + 12$

9 $6x^2 + 17x + 5$

10 $10x^2 - 21x - 10$

11 $12x^2 - 17x + 6$

12 $21x^2 + 19x - 12$

EXERCISE 28.6H (pt)

Simplify each of these algebraic fractions.

1 $\dfrac{x + 4}{x^2 + 2x - 8}$

2 $\dfrac{x^2 - 2x - 15}{x - 5}$

3 $\dfrac{x^2 - 7x + 12}{x^2 - x - 12}$

4 $\dfrac{x^2 - 2x - 8}{x^2 - 6x + 8}$

5 $\dfrac{3x^2 + 7x + 2}{x^2 - 7x + 6}$

6 $\dfrac{6x^2 + 8x - 8}{3x^2 + 7x - 6}$

7 $\dfrac{x^2 + x - 2}{x^2 - 1}$

8 $\dfrac{x^2 + x - 12}{x^2 - 16}$

9 $\dfrac{4x + 8}{x^2 + 6x + 8}$

10 $\dfrac{10x + 15}{4x^2 + 6x}$

11 $\dfrac{x^2 + 2x - 8}{x^2 - 5x + 6}$

12 $\dfrac{x^2 - 2x - 8}{x^2 - 16}$

13 $\dfrac{12x^2 + 15x}{4x^2 - 7x - 15}$

14 $\dfrac{3x^2 + 14x + 8}{6x^2 - 5x - 6}$

15 $\dfrac{3(x - 4)^2}{4x^2 - 64}$

16 $\dfrac{3x(x + 4)^2}{x^2 + x - 12}$

29 USING A CALCULATOR

EXERCISE 29.1H

Use your calculator to work out these.
Give all your answers to 3 significant figures.

1. **(a)** $\dfrac{1}{1.7} + \dfrac{1}{1.653}$

   (b) $\dfrac{1}{0.9} \times \dfrac{1}{8.24}$

   (c) $\dfrac{8.06}{5.91} - \dfrac{1.594}{1.62}$

2. **(a)** 0.741^3 **(b)** 2.28^{-5} **(c)** $(9.2 + 15.3)^2$

3. **(a)** $\sqrt[4]{12.2}$

   (b) $\sqrt[3]{0.8145 - 0.757}$

   (c) $\sqrt[5]{8.6^2 + 9.71^3}$

4. **(a)** $\sin 14.6°$

   (b) $\tan 71.3°$

   (c) $\sin 247° - \cos(-31)°$

5. **(a)** $\cos^{-1} 0.141$

   (b) $\sin^{-1} 0.464$

   (c) $\tan^{-1} \dfrac{1}{\sqrt{2}}$

6. **(a)** $(1.2 \times 10^4) \times (5.3 \times 10^6)$

   (b) $\dfrac{4.06 \times 10^{-2}}{7 \times 10^{-4}}$

   (c) $(5.9 \times 10^{-3}) \times (2.4 \times 10^{20})$

7. **(a)** $\dfrac{9.71 \times 0.008\,476\,5}{5.9^2}$

   (b) $\dfrac{81.7 + 1.52}{62.8}$

   (c) $\dfrac{9.61}{17.37 \times 224}$

8. **(a)** $\dfrac{101}{27.4 + 296}$

   (b) $\dfrac{3.14 \times 0.782}{22.4 - 15.5}$

   (c) $\dfrac{18.21 - 5.63}{23.48 + 19.76}$

9. **(a)** $3 \cos 12° - 5 \sin 12°$

   (b) $\dfrac{3.4 \times \sin 47.1°}{\sin 19.2°}$

   (c) $2.7^2 + 3.6^2 - 2 \times 2.7 \times 3.6 \cos 25°$

10. **(a)** $(3.6 \times 10^{-8})^2$

    (b) $\sqrt{4.84 \times 10^6}$

    (c) $\dfrac{(4.06 \times 10^6) + (1.15 \times 10^7)}{5.83 \times 10^{-6}}$

EXERCISE 29.2H (pt)

1. £30 000 is invested at 18% compound interest.
   (a) Write down a formula for the amount, a, the investment is worth after t years.
   (b) Calculate the value of the investment after
   (i) 5 years. **(ii)** 12 years.

2. The value of a currency, the scud, depreciates by 6% each month.
   I have 250 000 scuds.
   (a) Write down a formula for the value of the currency, c, after t months.
   (b) Calculate the value of 250 000 scuds after
   (i) 4 months. **(ii)** 9 months.
   (c) Use trial and improvement to work out how many months it will be before the 250 000 scuds are only worth as much as 100 000 scuds at the start.

3 The population of a rare species of animal is decaying exponentially following the formula

$$N = 60\,000 \times 2^{-t}$$

where N is the number of these animals present and t is the time in years.

(a) How many of these animals were there when the survey started?

(b) How many of these animals were there after
(i) 3 years?   (ii) 10 years?

(c) After how many years will this species of animal no longer exist; that is, after how many years will the number fall below 2?

4 Liz's employer tells her that her weekly wage will increase by a fixed percentage each year. They tell her that the formula they will use is

$$A = 500 \times 1.04^n.$$

(a) How much is Liz's current weekly wage?

(b) What is the rate of increase?

(c) What does the letter n stand for in the formula?

(d) How much will her wage be after
(i) 4 years?   (ii) 10 years?

5 The population of a country is increasing at a rate of 3% per year.

In 2005 the population was 38 million.

(a) Write down a formula for the population size, P, after t years.

(b) What will be the population in
(i) 2010?   (ii) 2100?

(c) How long will it take for the population to double from its 2005 size?

6 A sample of a radioactive element has a mass of 1 kg.

Its mass reduces by 10% each year.

(a) Write down a formula for the mass, m, of the element after t years.

(b) Calculate the mass after
(i) 4 years.   (ii) 8 years.

(c) Use trial and improvement to find how long it takes for the mass to halve.

7 (a) Draw the graph of $y = 3^x$ for values of x from 0 to 3.

(b) Use your graph to find
(i) the value of y when $x = 1.5$
(ii) the solution of the equation $3^x = 15$.

8 (a) Draw the graph of $y = 2^{-x}$ for values of x from -4 to 0.

(b) Use your graph to find
(i) the value of y when $x = -2.5$.
(ii) the solution of the equation $2^{-x} = 10$.

EXERCISE 29.3H

1 Find the upper and lower bounds for each of these measurements.

(a) The height of a house is 8.5 m, to the nearest 0.1 m.

(b) A child weighs 57 kg to the nearest kilogram.

(c) A door is 0.75 m wide to the nearest centimetre.

(d) The winning time for the 100 m race was 12.95 seconds to the nearest hundredth of a second.

(e) The volume of milk in a bottle is 500 ml to the nearest millilitre.

2 A garden is measured as 43 m to the nearest metre.

Write the possible length of the garden as an inequality.

3 A kitchen unit is 60 cm wide, to the nearest centimetre.

(a) Is it possible that the unit will fit into a gap 60 cm wide, to the nearest centimetre? Show how you decide.

(b) Is it certain that the unit will fit into the gap? Show how you decide.

4 Find the upper and lower bounds for each of these measurements.

(a) The total weight of 10 books, each book weighing 1.5 kg to the nearest 0.1 kg.

(b) The total length of 20 paper clips, each measuring 3 cm to the nearest centimetre.

(c) The total time to make 100 sandwiches when each sandwich takes 40 seconds to make, to the nearest second.

EXERCISE 29.4H pt

1 The lengths of the sides of a rectangle are
   8 cm and 10 cm.
   Both measurements are correct to the nearest
   centimetre.
   Work out the upper and lower bounds of the
   perimeter of the rectangle.

2 Work out the largest and smallest possible
   areas of a triangle with a base of 7 cm and a
   height of 5 cm, where both lengths are to the
   nearest centimetre.

3 Sugar weighing 0.1 kg is taken from a bag
   weighing 2 kg.
   Both weights are correct to the nearest 0.1 kg.
   What are the maximum and minimum possible
   weights of the sugar remaining?

4 Two stages of a relay race are run in times of
   14.07 seconds and 15.12 seconds, to the
   nearest 0.01 seconds.
   Calculate the upper bound of
   (a) the total time of the two stages.
   (b) the difference between the times for the
        two stages.

5 Given that $p = 5.1$ and $q = 8.6$, correct to
   1 decimal place, work out the largest and
   smallest possible values of
   (a) $p \times q$ **(b)** $q \div p$.

6 A train travels 150 miles in 1.8 hours.
   The distance is correct to the nearest mile and
   the time is correct to the nearest 0.1 hour.
   Work out the upper and lower bounds of the
   speed of the train.

7 The density of an object is given as 5.7 g/cm^3
   to the nearest 0.1 g/cm^3.
   Its volume is 72.5 cm^3 to 3 significant figures.
   Find the upper and lower bounds of the mass
   of the object.

8 Scott is trying to work out a value for π.
   He measures the circumference of a circle as
   32 cm and the diameter as 10 cm, both correct
   to the nearest centimetre.
   Calculate the upper and lower bounds of
   Scott's value for π.

9 Use the formula $a = \dfrac{v^2}{2s}$ to work out the
   upper and lower bounds of a when $v = 2.1$
   and $s = 5.7$ and both values are correct to 1
   decimal place.

10 Work out the upper and lower bounds of the
   following calculation.
   $$\frac{9.4 - 5.2}{3.8}$$
   Each value in the calculation is correct to
   2 significant figures.

EXERCISE 30.1H (pt)

Solve each of these equations.

1 $5(2x - 3) = 4x - 3$

2 $3(2x - 1) = 5(2x - 3)$

3 $2(4x - 1) = 6x + 3$

4 $\dfrac{x}{2} = 3x - 15$ **5** $\dfrac{x}{3} = 2x - 5$

6 $\dfrac{3x}{2} = 4 + x$ **7** $\dfrac{5x}{2} = 4x - 3$

8 $\dfrac{2x}{3} = x - \dfrac{1}{2}$ **9** $\dfrac{x}{2} = \dfrac{3x}{4} + \dfrac{1}{2}$

10 $\dfrac{x}{3} = \dfrac{3x}{4} - \dfrac{1}{12}$ **11** $\dfrac{3x}{2} = \dfrac{3x - 2}{5} + \dfrac{11}{5}$

12 $\dfrac{2x - 1}{6} = \dfrac{x - 3}{6} + \dfrac{1}{2}$

13 $\dfrac{x - 2}{3} - \dfrac{2x - 1}{2} = \dfrac{7}{6}$

14 $\dfrac{3x - 2}{2} + \dfrac{2x + 1}{6} + \dfrac{2}{9} = 0$

15 $\dfrac{4x - 3}{2} = \dfrac{x - 3}{3} + \dfrac{7}{6}$

EXERCISE 30.2H

Solve each of these equations.
Where the answer is not exact, give your answer correct to 3 significant figures.

1 $\dfrac{20}{x} = 40$ **2** $\dfrac{48}{x} = 12$ **3** $\dfrac{15}{2x} = 3$

4 $\dfrac{5}{3x} = \dfrac{1}{6}$ **5** $\dfrac{2}{3x} = \dfrac{4}{3}$ **6** $1.4x = 7.6$

7 $3.7x = 40$ **8** $\dfrac{x}{2.3} = 5.6$

9 $7.3(1.2x - 4.7) = 9.6$

10 $\dfrac{1.6}{x} = 2.7$

EXERCISE 30.3H (pt)

Solve each of these inequalities.

1 $5(x - 2) > 11 - 2x$

2 $4(3x - 4) \geqslant 3(2x - 1) + 2$

3 $\dfrac{5x}{2} < x + 6$ **4** $\dfrac{2x}{3} + 5 < 4x$

5 $\dfrac{x}{2} > \dfrac{3x}{4} + 2$ **6** $\dfrac{x}{5} \leqslant \dfrac{x}{4} - 2$

7 $3.5x < 4 + x$ **8** $14.6x \geqslant 7.1x + 4.9$

EXERCISE 30.4H (pt)

1 Draw a pair of axes and label them 0 to 6 for both x and y.
Show, by shading, the region where $x > 0$, $y > 0$ and $2x + 3y < 12$.

2 Draw a pair of axes and label them -2 to 2 for x and -4 to 8 for y.
Show, by shading, the region where $x < 2$, $y > -2$ and $y < 3x + 2$.

3 Draw a pair of axes and label them 0 to 3 for x and 0 to 9 for y.
Show, by shading, the region where $y < 6$, $y < 3x$ and $y > 2x$.

4 Draw a pair of axes and label them 0 to 8 for both x and y.
Show, by shading, the region where $y > 0$, $y < x$ and $3x + 4y < 24$.

5 Draw a pair of axes and label them 0 to 8 for both x and y.
Show, by shading, the region where $y > 0$, $8x + 3y < 24$ and $5x + 6y > 30$.

EXERCISE 31.1H

1 Calculate the length of the diagonal of a cuboid measuring 6 cm by 10 cm by 5 cm.

2 The length of the diagonal of a cube is 6.8 cm. Find the length of a side of this cube.

3 In this cuboid, AB = 10 cm, BC = 6 cm and CG = 8 cm.

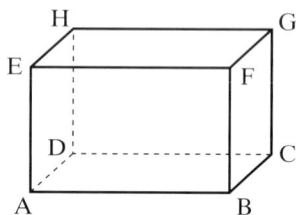

Calculate
(a) angle GDC.
(b) the length of EG.
(c) the length of HB.
(d) angle BHD.

4 A pyramid is 8 cm high and has a square base of side 6 cm.
Its sloping edges are all of equal length.
Calculate the length of a sloping edge.

5 The length of each sloping edge of a square-based pyramid is 12 cm.
The sides of the base are each 10 cm.
Calculate the height of the pyramid.

6 Calculate the distance between the points (4, −3, 7) and (−7, −6, 2).

7 ABCDEF is a triangular wedge.
The faces ABFE, BCDF and ACDE are rectangles.

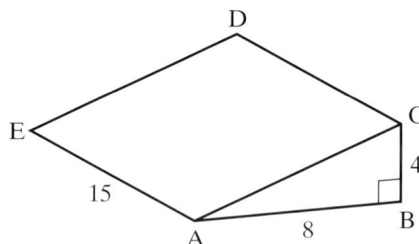

(a) Calculate the length AD.
(b) Calculate angle DAC.

8 A is 30 m due south of a church tower.
From A, the elevation of the top of the tower is 51°.
(a) Calculate the height of the tower.
From B, which is due west of the tower, the angle of elevation of the top of the tower is 35°.
(b) Calculate how far B is from the tower.
(c) Calculate the distance AB, assuming that A and B are on the same level as the base of the tower.

9 A pyramid VABCD has a square base ABCD of side 6 cm.
O is the centre of the base.
(a) Show that AO = $\sqrt{18}$ cm.
Angle VAO, the angle between a sloping edge and the base, is 62°.
(b) Calculate the height VO.
(c) Calculate the length of a sloping edge of the pyramid.

10 The pyramid OABCD has a horizontal rectangular base ABCD as shown.
O is vertically above A.

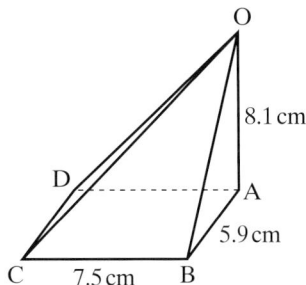

Calculate
(a) the length of OB.
(b) angle OCB.
(c) the length of OC.

EXERCISE 31.2H

1 A square-based pyramid has sloping edges of length 9.5 cm.
The sloping edges make an angle of 60° with the base.
Calculate the height of the pyramid.

2 A pyramid is 7 cm high and has a square base of side 5 cm.
Its sloping edges are all of equal length.
Calculate the angle between a sloping face and the base.

3 In this cuboid, AB = 12 cm, BC = 5 cm and CG = 6 cm.

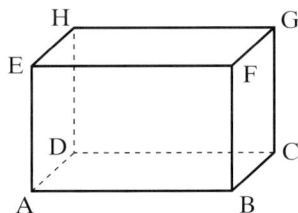

Calculate
(a) the length of AC.
(b) the length of AG.
(c) the angle between AG and the base ABCD.
(d) the angle between AG and face BCGF.

4 The length of the diagonal of a cuboid is 9.3 cm.
The height of the cuboid is 5.6 cm.
Calculate the angle between the diagonal and a vertical edge of the cuboid.

5 VABCD is a square-based pyramid of height 8 cm.
Its base ABCD has side 10 cm.
All its sloping edges are equal in length.
M is the midpoint of AB.

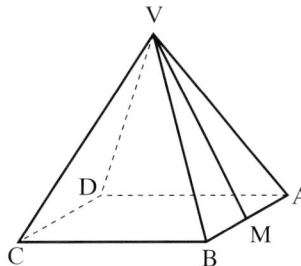

Calculate
(a) the angle which VM makes with the base.
(b) the length of VM.
(c) the length of VA.
(d) the angle which VA makes with the base.

6 VABCD is a square-based pyramid of height 8 cm.
Its base ABCD has side 10 cm.
V is directly above A.

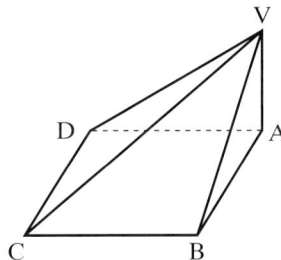

Calculate the lengths of the sloping edges VB, VC and VD, and the angles they make with the base.

7 This 'lean-to' workshop is a prism with a trapezium as its cross-section.

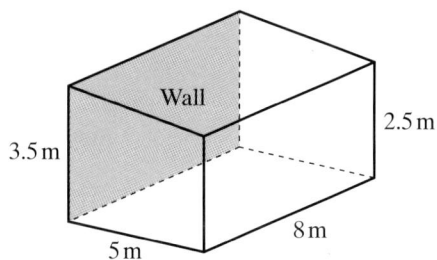

(a) Calculate the area of the sloping roof.
(b) Calculate the angle between the roof and the wall against which the workshop is built.
(c) Calculate the length of the longest piece of wood that will fit in the workshop. (Assume that the door is large enough for it to get in!)

8 The diagonal of a cuboid has length 12.4 cm. It makes an angle of 33° with the base of the cuboid.
(a) Calculate the height of the cuboid.
(b) The length of the base of the cuboid is 5.8 cm. Calculate its width.

EXERCISE 32.1H (pt)

Rearrange each of these formulae to make the letter in brackets the subject.

1   $C = 2\pi r$ (r)

2   $k = \dfrac{PV}{T}$ (V)

3   $ax + b = 2x + 3b$ (x)

4   $a(x - y) = 3(x + y)$ (x)

5   $pq - r = rq - t$ (p)

6   $pq - r = rq - t$ (q)

7   $pq - r = rq - t$ (r)

8   $s = \frac{1}{2}(u + v)t$ (t)

9   $s = \frac{1}{2}(u + v)t$ (v)

10   $V = \dfrac{1}{x} - \dfrac{1}{3}$ (x)

EXERCISE 32.2H (pt)

1   Rearrange each of these formulae to make the letter in brackets the subject.

(a) $y = 2x^2 + 3$ (x)

(b) $h = \dfrac{gt^2}{4\pi^2}$ (t)

(c) $y = \sqrt{\dfrac{x}{3}}$ (x)

(d) $y = \sqrt{x + a}$ (x)

(e) $s = \sqrt{x^2 + y^2}$ (x)

(f) $A = \frac{1}{3}\pi r (l + 3r)$ (l)

(g) $y = \dfrac{2x}{3} - 5$ (x)

(h) $y = (x - a)^2$ (x)

2   The formula for the volume of a sphere is $V = \frac{4}{3}\pi r^3$ where r is the radius of the sphere.
   (a) Find the volume of a sphere of radius 3 cm.
        Give your answer to 1 decimal place.
   (b) Rearrange the formula to make r the subject.
   (c) What is the radius of a sphere with volume 500 cm³?
        Give your answer to 1 decimal place.

3   The formula for the geometric mean, m, of two numbers a and b is $m = \sqrt{ab}$.
   For three numbers a, b and c the geometric mean is $m = \sqrt[3]{abc}$, and for four numbers a, b, c and d it is $\sqrt[4]{abcd}$.
   (a) Use the appropriate formula to find the geometric mean of the numbers 5, 8 and 12. Give your answer correct to 2 decimal places.
   (b) The geometric mean of four numbers 4, 7, 8 and x is 6.88, correct to 2 decimal places.
        (i)   Rearrange the appropriate formula to make x the subject.
        (ii) Find x correct to the nearest whole number.

33 → PROPORTION AND VARIATION

EXERCISE 33.1H

1 A carousel revolves 14 times in 252 seconds. How many revolutions does the carousel make in 90 seconds?

2 An express train travels 63 miles in 35 minutes. How long would it take the train to travel 45 miles at the same speed?

3 Sound can travel 5145 metres in 15 seconds. How far can sound travel in 24 seconds?

4 A piece of wax with a volume of 240 cm^3 has a mass of 216 grams.
What is the mass of 1000 cm^3 of wax?

5 Tiling a floor with an area of 15 m^2 uses 21 litres of adhesive.
How much adhesive would be needed to tile a floor with an area of 35 m^2?

6 A 45 minute telephone call costs £1.80. How much does it cost to make a 12 minute call at the same rate?

7 A contractor is paid £75 for working 6 hours. How much would the contractor be paid for working 10 hours at the same rate?

8 A machine cleans a carpet with an area of 12 m^2 using 21 litres of water.
How much water would the machine use to clean a similar carpet with an area of 40 m^2?

9 An aeroplane travels 54 km in 6 minutes. How far would it travel in 15 minutes at the same speed?

10 A person can walk 340 metres in 4 minutes. How far would the same person walk in 3 minutes?

EXERCISE 33.2H pt

1 A journey takes 30 minutes at a constant speed of 40 miles per hour. How long would the journey take at a constant speed of 60 miles per hour?

2 It takes a team of 12 men 10 weeks to lay a pipeline. How long would the pipeline take to lay if there were 15 men?

3 A pool can be emptied in 18 hours using four pumps. How long will it take to empty the pool using three pumps?

4 A supply of hay is enough to feed eight horses for 30 days. How long would the same supply feed 20 horses?

5 Using three ploughs it is possible to plough a field in 6 hours. How long would it take to plough the same field using two ploughs?

6 It takes a team of three men 14 hours to install a fence. How long would it take to install the fence if there were eight men?

7 A tank can be filled using three pumps in a period of 28 hours. How long would it take to fill the tank using seven pumps?

8 A carpet with an area of 16 m^2 costs £440. What is the cost of 26 m^2 of the same carpet?

9 A supply of corn is enough to feed 24 pigs for 35 days. How long would the same supply of corn feed 60 pigs?

10 An express train completes a journey of 473 miles in 5.5 hours. How long would it take the train to travel 129 miles at the same speed?

EXERCISE 33.3H

For each of these relationships
(a) state the type of proportion.
(b) find the formula.
(c) find the missing y value in the table.

1

x	1	8	15
y	6	48	

2

x	4	12	18
y	36	12	

3

x	4	10	36
y	2	12.5	

4

x	3	4	5
y	8	4.5	

5

x	2	5	20
y	30	12	

6

x	4	60	100
y	5	72	

7

x	4	7	40
y	35	20	

8

x	2	3	6
y	9	4	

9

x	1	4	10
y	10	160	

10

x	3	15	75
y	15	75	

11

x	3	4	10
y	24	18	

12

x	2	5	32
y	8	50	

13

x	12	60	150
y	4	20	

14

x	5	10	20
y	8	2	

EXERCISE 33.4H

For each of these relationships
(a) state the type of proportion.
(b) find the formula.

1

x	2	6
y	8	72

2

x	1	3
y	36	4

3

x	5	12
y	8.64	1.5

4

x	6	10
y	9	25

5

x	1	5
y	100	4

EXERCISE 34.1H

Find the equation of each of these straight lines.

1

2

3

4

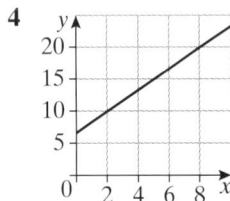

5 A line with gradient -2, passing through the point $(2, 0)$.

6 A line with gradient $\frac{3}{4}$, passing through the point $(4, 2)$.

7 A line passing through $(1, 3)$ and $(4, 9)$.

8 A line passing through $(2, 3)$ and $(5, -6)$.

9 A line passing through $(-1, 5)$ and $(3, -5)$.

10 A line passing through $(3, 1)$ and $(-3, -7)$.

EXERCISE 34.2H pt

1 Find the gradient of a line perpendicular to the line joining each of these pairs of points.
  (a) $(1, 1)$ and $(5, 3)$
  (b) $(1, 2)$ and $(4, -2)$
  (c) $(-1, 5)$ and $(2, 8)$

2 Find the equation of the line that passes through $(1, 0)$ and is parallel to $y = 2x + 6$.

3 Find the equation of the line that passes through $(2, 3)$ and is parallel to $4x + 2y = 7$.

4 **(a)** State the gradient of the line $3x + 5y = 6$.
  (b) Find the equation of the line perpendicular to $3x + 5y = 6$ that passes through $(1, 1)$.

5 Find the equation of the line that passes through $(4, 1)$ and is perpendicular to $y = 4x + 3$.

6 Find the equation of the line that passes through $(0, 5)$ and is perpendicular to $3y = x - 1$.

7 Find the equation of the line that passes through $(2, 5)$ and is perpendicular to $7y + 2x = 9$.

8 Which of these lines are
  (a) parallel?   **(b)** perpendicular?
  $y = x + 5$ $y = 3x + 5$
  $x + 3y = 5$ $4x - y = 5$

9 Two lines cross at right angles at the point $(2, 5)$. One passes through $(4, 7)$. What is the equation of the other line?

10 In the diagram AC is a diagonal of the square ABCD. Work out
  (a) the equation of the line AC.
  (b) the equation of the line BD.
  (c) the coordinates of B and D.

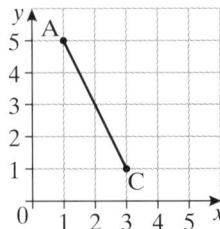

EXERCISE 35.1H 🅟🅣

1 Triangles ABC and PQR are congruent.

    (a) Write down the size of angle
        (i) BAC
        (ii) BCA
        (iii) PQR
        (iv) RPQ
    (b) Which side in triangle PQR is 7.5 cm long?

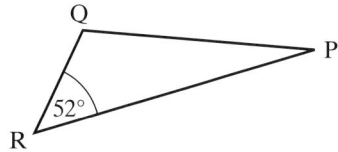

2 Which other triangles are congruent to triangle ABC?

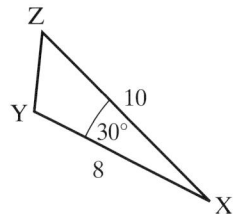

3 'SAS' is shorthand for 'Two sides and the included angle of one triangle are equal to the two sides and the included angle of the other triangle.'
Explain what each of these is shorthand for.
    (a) SSS   **(b)** ASA   **(c)** RHS

4 State whether or not each of these pairs of triangles are congruent.
Give reasons for your answers.

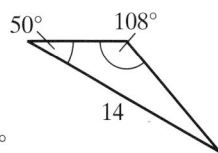

5 In the diagram, AB = BC and angle ABD = angle DBC.
Prove that triangle ABD is congruent to triangle CBD.

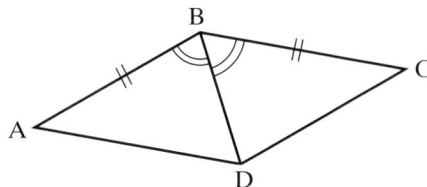

6 In the diagram, AB is parallel to ED and BC = CE.
Prove that triangles ABC and DEC are congruent.

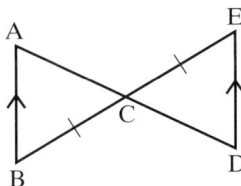

7 **(a)** Sketch a quadrilateral ABCD in which angle ABC = angle ADC and BC is parallel to AD.
Join the vertices A and C to make two triangles.
(You may recognise the shape you have drawn, but do not make any other assumptions about the shape except what you have been told.)

(b) Prove that triangles ABC and CDA are congruent.

(c) What does this prove about the opposite sides of the quadrilateral?

8 **(a)** Sketch an isosceles triangle ABC with AB = AC.
Draw a straight line from the midpoint of BC to A.

(b) Use congruent triangles to prove that this line bisects the angle at A and is perpendicular to the side BC.

9 **(a)** Sketch an equilateral triangle ABC.
The midpoint of AB is X, the midpoint of BC is Y, and the midpoint of AC is Z.

(b) Prove that triangle XYZ is also equilateral.

10 Two triangles have two sides of length 5 cm and 9 cm.
The angle opposite the 5 cm side is 25°.
The two triangles are not congruent.
Sketch the two triangles to show that they are not congruent.

36 → QUADRATIC EQUATIONS

EXERCISE 36.1H

Solve each of these quadratic equations.

1 $x(x + 3) = 0$

2 $x(x - 4) = 0$

3 $(x - 2)(x + 2) = 0$

4 $(x - 5)(x + 6) = 0$

5 $3x(x + 5) = 0$

6 $(x + 10)(x - 1) = 0$

7 $(x - 7)(x - 3) = 0$

8 $(x + 9)(2x - 5) = 0$

9 $(x - 3)(4x - 1) = 0$

10 $(3x - 7)(2x + 1) = 0$

11 $4x(3x - 2) = 0$

12 $(2x - 3)(3x - 4) = 0$

EXERCISE 36.2H

Solve each of these quadratic equations.

1 $x^2 + 5x + 4 = 0$

2 $x^2 - 8x + 7 = 0$

3 $x^2 + 4x - 5 = 0$

4 $x^2 - x - 2 = 0$

5 $x^2 + x - 12 = 0$

6 $x^2 + 7x = 0$

7 $x^2 - 9x + 14 = 0$

8 $x^2 - 3x - 10 = 0$

9 $3x^2 - 15x = 0$

10 $x^2 - 16 = 0$

11 $x^2 + 8x + 12 = 0$

12 $x^2 - 8x - 20 = 0$

13 $x^2 + 8x + 16 = 0$

14 $x^2 - 8x + 15 = 0$

15 $x^2 - 100 = 0$

16 $x^2 + 21x - 22 = 0$

17 $2x^2 - 6x = 0$

18 $x^2 - 11x + 18 = 0$

19 $4x^2 + 2x = 0$

20 $9x^2 - 25 = 0$

21 $2x^2 + 3x + 1 = 0$

22 $3x^2 - 5x + 2 = 0$

23 $3x^2 + 10x + 7 = 0$

24 $2x^2 + 5x - 3 = 0$

25 $3x^2 + 4x - 4 = 0$

26 $2x^2 - 17x + 8 = 0$

27 $4x^2 - 4x + 1 = 0$

28 $4x^2 - 11x - 3 = 0$

29 $2x^2 - x - 10 = 0$

30 $6x^2 - 7x - 5 = 0$

EXERCISE 36.3H

Solve each of these quadratic equations.

1 $x^2 + x = 6$

2 $x^2 = 7x - 10$

3 $x^2 = 4x + 5$

4 $x^2 = 10x - 21$

5 $x^2 = 11x$

6 $x^2 = 12 - 4x$

7 $x^2 = 6 + 5x$

8 $2x^2 = 3x - 1$

9 $9 + 8x - x^2 = 0$

10 $5 - 4x - x^2 = 0$

EXERCISE 36.4H

1 (a) Write each of these quadratic expressions
       in the form $(x + m)^2 + n$.
       (i) $x^2 + 2x$ **(ii)** $x^2 - 4x$
       (iii) $x^2 - 14x$ **(iv)** $x^2 - x$

   (b) Write each of these quadratic expressions
       in the form $(x + m)^2 + n$.
       Use your answers to part **(a)**.
       (i) $x^2 + 2x - 5$ **(ii)** $x^2 - 4x + 7$
       (iii) $x^2 - 14x + 1$ **(iv)** $x^2 - x - 7$

2 For each of these quadratic expressions,
   complete the square; that is, write it in the
   form $(x + m)^2 + n$.
   (a) $x^2 + 6x - 1$
   (b) $x^2 + 8x - 2$
   (c) $x^2 - 4x + 3$
   (d) $x^2 - 2x - 3$
   (e) $x^2 - 12x + 37$
   (f) $x^2 + 10x - 3$
   (g) $x^2 - 6x + 19$
   (h) $x^2 + 3x - 2$
   (i) $x^2 - 5x + 7$

EXERCISE 36.5H

In this exercise, give all your answers correct to
2 decimal places.

1 Solve each of these quadratic equations.
   (a) $(x - 2)^2 - 5 = 0$
   (b) $(x + 3)^2 - 7 = 0$
   (c) $(x - 4)^2 - 20 = 0$
   (d) $(x + 1)^2 - 11 = 0$

2 For each of these quadratic equations, first
   complete the square and then solve the
   equation.
   (a) $x^2 - 6x - 5 = 0$
   (b) $x^2 + 4x + 1 = 0$
   (c) $x^2 - 4x - 3 = 0$
   (d) $x^2 + 10x + 5 = 0$
   (e) $x^2 + 2x - 8 = 0$
   (f) $x^2 + 6x - 2 = 0$
   (g) $x^2 - 8x + 6 = 0$
   (h) $x^2 + 14x - 3 = 0$
   (i) $x^2 + 3x - 6 = 0$

EXERCISE 36.6H

Solve each of these quadratic equations using the
formula.
Give your answers correct to 2 decimal places.
If there are no real solutions, say so.

1 $x^2 + 3x + 1 = 0$ 2 $x^2 - 7x + 3 = 0$

3 $x^2 + 2x - 11 = 0$ 4 $x^2 + x - 7 = 0$

5 $2x^2 + 6x + 3 = 0$ 6 $2x^2 + 5x + 1 = 0$

7 $x^2 + 4x + 7 = 0$ 8 $3x^2 + 10x + 5 = 0$

9 $3x^2 - 7x - 4 = 0$ 10 $2x^2 - x - 8 = 0$

11 $5x^2 + 8x + 1 = 0$ 12 $2x^2 + 5x - 7 = 0$

13 $4x^2 - 12x + 6 = 0$ 14 $2x^2 - 5x - 4 = 0$

15 $2x^2 + 10x + 15 = 0$ 16 $3x^2 + 3x - 2 = 0$

17 $x^2 - 3x - 50 = 0$ 18 $4x^2 + 9x + 3 = 0$

19 $5x^2 + 11x + 4 = 0$ 20 $3x^2 - x - 8 = 0$

21 $2x^2 + 6x + 5 = 0$

EXERCISE 37.1H

Solve each of these pairs of simultaneous equations.

1 $2x + y = 8$
$3x - y = 7$

2 $4x + 5y = 3$
$3x + 5y = 1$

3 $2x + 5y = 6$
$2x + 3y = 14$

4 $4x - 5y = 7$
$4x + y = -11$

5 $2x + 3y = 9$
$6x - 3y = 39$

6 $3x + 5y = 23$
$7x + 5y = 37$

7 $5x + 2y = 5$
$-5x - 3y = 5$

8 $x + 3y = 13$
$2x + y = 11$

9 $3x + 2y = 16$
$5x + 4y = 27$

10 $4x - 3y = -3$
$2x + 5y = 18$

11 $3x \quad 2y - 19$
$x + 4y = -3$

12 $4x - 3y = 1$
$x + y = -5$

13 $3x + 2y = 4$
$x + 4y = 13$

14 $5x + y = 10$
$3x - 5y = 10$

15 $7x + 3y = 0$
$2x - 9y = 69$

16 $3x + 4y = 12$
$6x - 3y = 2$

17 $4x - 3y = 11$
$5x - y = 22$

18 $3x + 4y = 19$
$5x + 3y = 17$

19 $3x - 2y = 13$
$7x + 6y = 9$

20 $6x - 5y = -16$
$5x - 3y = -11$

21 $4x + 2y = 17$
$3x + 5y = 18$

EXERCISE 37.2H (pt)

1 Solve each of these pairs of simultaneous equations by substitution.

(a) $y = 2x - 10$
$y = 3x - 13$

(b) $y = 5x + 18$
$y = 4 - 2x$

(c) $y = 3x + 11$
$x + y = 3$

(d) $y = 8 - 2x$
$2x + 5y = 48$

(e) $x = 3y - 5$
$3x - 2y = 6$

(f) $x - 2y = 7$
$y = 3x + 4$

2 Find, graphically, the coordinates of the points where the circle $x^2 + y^2 = 100$ and the line $y = x + 2$ intersect.

3 Use substitution to find the coordinates of the points where each of these circles and straight lines intersect.
(a) $x^2 + y^2 = 65$
$y = x + 3$
(b) $x^2 + y^2 = 53$
$y = x - 5$

4 Calculate the radius of each of these circles. Give your answers correct to two decimal places.
(a) $x^2 + y^2 = 65$ **(b)** $x^2 + y^2 = 53$

5 Find the coordinates of the points where the line $y = 6 - x$ intersects the curve $y = 2x^2 - 4$.

38 → DECIMALS AND SURDS

EXERCISE 38.1H (pt)

1 Which of these fractions are equivalent to recurring decimals?
   (a) $\frac{1}{8}$ (b) $\frac{7}{30}$ (c) $\frac{4}{7}$
   (d) $\frac{5}{16}$ (e) $\frac{11}{120}$

2 Find the decimal equivalent of each of the fractions in question **1**.

3 When these fractions are written as decimals, which of them terminate?
   (a) $\frac{6}{40}$ (b) $\frac{7}{8}$ (c) $\frac{1}{6}$
   (d) $\frac{2}{75}$ (e) $\frac{11}{80}$

4 Find the decimal equivalent of each of the fractions in question **3**.

5 Find the fractional equivalent of each of these terminating decimals.
   Write each fraction in its simplest form.
   (a) 0.16   (b) 0.305
   (c) 0.625   (d) 0.408

6 Find the fractional equivalent of each of these recurring decimals.
   Write each fraction in its simplest form.
   (a) $0.\dot{3}$ (b) $0.\dot{8}$ (c) $0.1\dot{5}$ (d) $0.7\dot{2}$

7 Find the fractional equivalent of each of these recurring decimals.
   Write each fraction in its simplest form.
   (a) $0.\dot{6}\dot{3}$ (b) $0.\dot{4}\dot{7}$ (c) $0.1\dot{5}$ (d) $0.3\dot{8}$

8 Find the fractional equivalent of each of these recurring decimals.
   Write each fraction in its simplest form.
   (a) $0.\dot{3}0\dot{6}$ (b) $0.\dot{4}1\dot{4}$
   (c) $0.0\dot{8}$ (d) $0.0\dot{2}\dot{7}$

EXERCISE 38.2H

1 Simplify these.
   (a) $\sqrt{7} + 3\sqrt{7}$
   (b) $9\sqrt{5} - 2\sqrt{5}$
   (c) $4\sqrt{3} - \sqrt{3}$
   (d) $\sqrt{3} \times 5\sqrt{3}$
   (e) $6\sqrt{2} \times \sqrt{5}$
   (f) $3\sqrt{3} \times \sqrt{12}$

2 Write each of these expressions in the form $a\sqrt{b}$, where b is an integer which is as small as possible.
   (a) $\sqrt{18}$
   (b) $3\sqrt{50}$
   (c) $5\sqrt{12}$
   (d) $6\sqrt{20}$
   (e) $\sqrt{108}$
   (f) $\sqrt{1250}$

3 Simplify these.
   (a) $\sqrt{18} + 4\sqrt{2}$
   (b) $7\sqrt{3} - \sqrt{75}$
   (c) $\sqrt{12} + 3\sqrt{3}$
   (d) $\sqrt{8} \times 3\sqrt{6}$
   (e) $\sqrt{21} \times \sqrt{12}$
   (f) $2\sqrt{250} \times \sqrt{10}$

4 Expand and simplify these.
   (a) $\sqrt{2}(5 + \sqrt{2})$
   (b) $\sqrt{3}(\sqrt{12} + \sqrt{2})$
   (c) $4\sqrt{5}(2 + \sqrt{45})$
   (d) $3\sqrt{2}(\sqrt{18} + \sqrt{5})$

5 Expand and simplify these.

    (a) $(1 + 2\sqrt{7})(3 + 4\sqrt{7})$

    (b) $(3 - \sqrt{3})(8 + \sqrt{3})$

    (c) $(4 - \sqrt{5})(3 - 2\sqrt{5})$

    (d) $(\sqrt{5} + 1)(\sqrt{5} - 4)$

    (e) $(\sqrt{10} + 5)(\sqrt{10} - 5)$

    (f) $(6 - \sqrt{13})(4 - 2\sqrt{13})$

6 Find the value of each of these expressions, when $m = 4 + \sqrt{5}$ and $n = 6 - 2\sqrt{5}$.

    (a) $3n$ **(b)** $m + 2n$

    (c) $3m - 2n$ **(d)** mn

7 Find the value of each of these expressions, when $p = 7 + 2\sqrt{3}$ and $q = 7 - 2\sqrt{3}$.

    (a) $4p$ **(b)** $\sqrt{3}q$

    (c) $p - q$ **(d)** pq

    (e) p^2 **(f)** q^2

8 $r = 7 + \sqrt{5}$ and $s = 6 - 4\sqrt{5}$.

    (a) Find the value of k if $3r + ks$ is an integer.

    (b) Find the value of k if $3r + ks$ is of the form $a\sqrt{5}$.

9 Rationalise the denominator and simplify each of these.

    (a) $\dfrac{15}{\sqrt{3}}$ **(b)** $\dfrac{4}{\sqrt{5}}$

    (c) $\dfrac{6}{5\sqrt{2}}$ **(d)** $\dfrac{6}{7\sqrt{12}}$

    (e) $\dfrac{6\sqrt{3}}{5\sqrt{2}}$ **(f)** $\dfrac{12\sqrt{10}}{11\sqrt{6}}$

10 Rationalise the denominator and simplify each of these.

    (a) $\dfrac{10 + 2\sqrt{5}}{\sqrt{5}}$

    (b) $\dfrac{12 + \sqrt{3}}{4\sqrt{3}}$

    (c) $\dfrac{6 + 5\sqrt{2}}{\sqrt{2}}$

    (d) $\dfrac{5 + 10\sqrt{3}}{\sqrt{15}}$

EXERCISE 39.1H

1 This is a distance–time graph for a particle moving in a straight line.

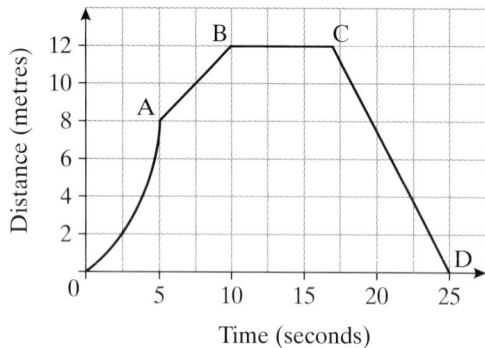

Describe the motion of the particle.

2 (a) Draw a distance–time graph with the Time axis (t) from 0 to 10 seconds and the Distance axis (s) from 0 to 20 metres. Draw a straight line joining $(0, 0)$ to $(8, 18)$.
   (b) Calculate the velocity and describe the motion.

3 (a) Draw a velocity–time graph with the Time axis (t) from 0 to 10 seconds and the Velocity axis (v) from 0 to 20 m/s. Show a constant acceleration from $t = 0$, $v = 0$ to $t = 10$, $v = 15$.
   (b) What is the velocity when $t = 6$?
   (c) What is the acceleration?

4 A particle moves with a constant acceleration of 2 m/s^2 from $t = 0$ to $t = 3$.
   It moves at a constant velocity for the next 5 seconds.
   Then it moves with a constant acceleration of -1.25 m/s^2 for the next 12 seconds.
   (a) Draw a velocity–time graph with the Time axis (t) from 0 to 20 seconds and the Velocity axis (v) from -10 to $+10$ m/s. Show the movement of the particle.
   (b) What is the constant velocity from $t = 3$ to $t = 8$?
   (c) When is the velocity zero?
   (d) What is the velocity when $t = 20$?

5 This is a velocity–time graph to show the movement of a particle.

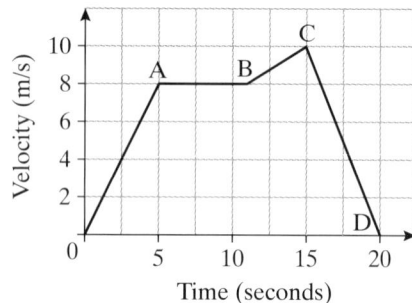

Describe as fully as possible the movement of the particle.
Work out the accelerations where necessary.

EXERCISE 39.2H

1   (a) Draw the graph of $y = x^2 - 3x + 2$ for values of x from -1 to 4.
    (b) On the same axes, draw the line $y = x - 1$.
    (c) Write down the coordinates of the points where the line and the curve intersect.

2   (a) Draw the graph of $y = x^2 - 2x - 3$ for values of x from -2 to 4.
    (b) On the same axes, draw the line $y = 3 - x$.
    (c) Write down the coordinates of the points where the line and the curve intersect.

3   (a) Draw the graph of $y = x^2 - 6x + 4$ for values of x from 0 to 6.
    (b) On the same axes, draw the line $4y = 3x - 12$.
    (c) Write down the coordinates of the points where the line and the curve intersect.

4   (a) Draw the graph of $y = 10 + x - 2x^2$ for values of x from -3 to 3.
    (b) On the same axes, draw the line $3y + 4x = 12$.
    (c) Write down the coordinates of the points where the line and the curve intersect.

EXERCISE 39.3H pt

1   (a) Draw the graph of $y = x^2 - 5x - 6$ for values of x from -2 to 7.
    (b) Use your graph to solve these equations.
         (i) $x^2 - 5x - 6 = 0$
         (ii) $x^2 - 5x - 6 = -10$
         (iii) $x^2 - 5x - 6 = 2 - 3x$

2   (a) Draw the graph of $y = x^2 - 4x + 3$ for values of x from -1 to 5.
    (b) Use your graph to solve these equations. Give your answers to 1 decimal place.
         (i) $x^2 - 4x + 3 = 0$
         (ii) $x^2 - 4x - 3 = 0$
         (iii) $x^2 - 6x + 7 = 0$

3   (a) Draw the graph of $y = x^2 + 2x - 15$ for values of x from -6 to 4.
    (b) Use your graph to solve these equations. Give your answers to 1 decimal place.
         (i) $x^2 + 2x - 15 = 0$
         (ii) $x^2 + 2x - 10 = 0$
         (iii) $x^2 + 3x - 4 = 0$
    (c) (i) What line would you need to draw on the graph to solve the equation $x^2 + x + 3 = 0$?
         (ii) Why does this not work?

For the remaining questions, do not draw the graphs.

4   The graph of $y = x^2 - 2x + 1$ has been drawn. What other line needs to be drawn to solve each of these equations?
    (a) $x^2 - 2x + 1 = 5$
    (b) $x^2 - 2x + 1 = 3x - 2$

5   The graph of $y = x^2 - x - 12$ has been drawn. What other line needs to be drawn to solve each of these equations?
    (a) $x^2 - 3x - 12 = 0$
    (b) $x^2 - x = 0$
    (c) $x^2 + x - 15 = 0$

6   The graph of $y = x^2 - 6x + 8$ has been drawn. What other line needs to be drawn to solve each of these equations?
    (a) $x^2 - 6x + 4 = 0$
    (b) $x^2 - 8x + 8 = 0$
    (c) $x^2 - 4x + 3 = 0$

EXERCISE 39.4H

1 (a) Copy and complete this table of values for
       the equation $y = \dfrac{12}{x}$.

x	1	2	3	4	6	8	12
y							

   (b) Draw the graph of $y = \dfrac{12}{x}$ for values of x
        between 1 and 12.
   (c) Use your graph to estimate the value of y
        when $x = 9$.
   (d) (i) Find the point on the graph where
             $x = y$.
        (ii) Explain the connection of this
             number with 12.

2 (a) Draw the graph of $y = 2x^3$ for values of x
        between -3 and 3.
   (b) Draw the line $y = 10x$ on the same graph.
   (c) (i) Find the equation that is given at
             the intersection of the line and the
             curve.
        (ii) Estimate the solutions to this
             equation.

3 (a) Draw the graph of $y = \dfrac{6}{x + 1}$, for values
        of x between 0 and 5.
   (b) Use your graph to find the value of x
        when $y = 4$.
   (c) Draw the line $y = x - 1$ on the same
        graph.
   (d) (i) Find the equation that is given at the
             intersection of the line and the curve.
        (ii) Estimate the solutions to this equation.

4 (a) Draw a graph of $y = 4^{-x}$ for values of x
        between -2 and 2.
   (b) Use your graph to estimate
        (i) the value of y when $x = 0.5$.
        (ii) the solution to the equation $4^{-x} = 10$.

5 The table shows two pairs of values for the
   equation $y = ab^x$.

x	0	1	2	3	4	5
y	4	8				

   (a) Find the values of a and b.
   (b) Copy and complete the table.
   (c) Draw the graph of $y = ab^x$ for values of x
        from 0 to 5.
   (d) Use your graph to estimate the value of x
        when $y = 80$.

40 → TRIGONOMETRY 2

EXERCISE 40.1H

1 Find the area of each of these triangles.

(a)

(b)

(c) P

(d)

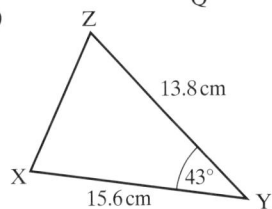

2 (a) In triangle ABC, $b = 9$ cm, $c = 16$ cm and the area is 60.4 cm^2. Find the size of angle A.
(b) In triangle PQR, $q = 6.4$ cm, $r = 7.8$ cm and the area is 12.1 cm^2. Find the size of angle P.
(c) In triangle XYZ, $x = 23.7$ cm, $y = 16.3$ cm and the area is 184 cm^2. Find the size of angle Z.

3 (a) In triangle ABC, $b = 20$ cm, angle C is 27° and the area is 145.3 cm^2.
Find the length of side BC.

(b) In triangle XYZ, $x = 9.3$ cm, angle Z is 94° and the area is 34.3 cm^2.
Find the length of side XZ.

EXERCISE 40.2H pt

1 Find the size of each of the sides and angles not given in these diagrams.

(a)

(b) R

(c)

(d) A

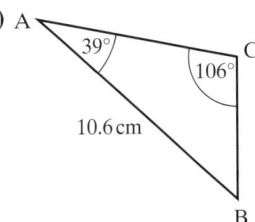

2 In triangle PQR, angle PQR is 38°, side PR is 8.3 cm and side PQ is 12 cm.
Calculate the size of the largest angle of the triangle.

EXERCISE 40.3H pt

1 Find the size of each of the sides and angles marked in these diagrams.

(a)

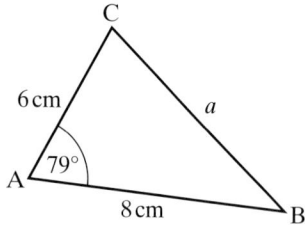

6 cm
a
79°
A
8 cm
B
C

(b)

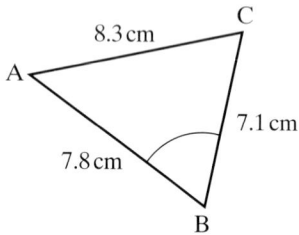

8.3 cm
C
A
7.1 cm
7.8 cm
B

(c)

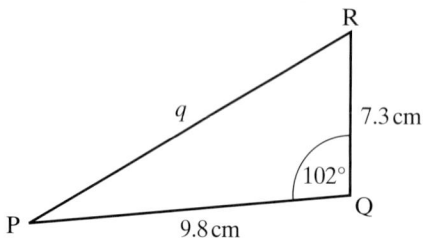

R
q
7.3 cm
102°
Q
P
9.8 cm

(d) X

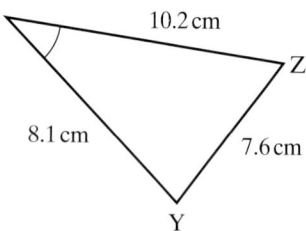

10.2 cm
Z
8.1 cm
7.6 cm
Y

2 Find the size of each of the angles in these diagrams.

(a)

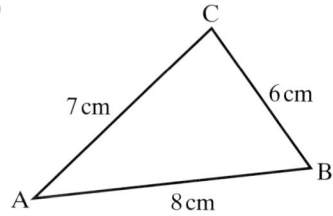

C
7 cm
6 cm
A
8 cm
B

(b)

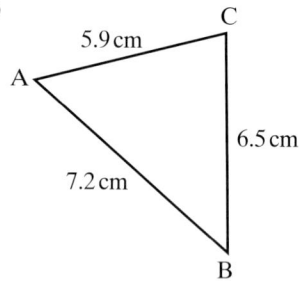

5.9 cm
C
A
6.5 cm
7.2 cm
B

(c)

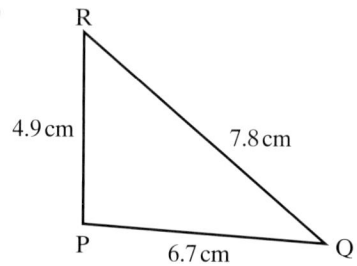

R
4.9 cm
7.8 cm
P
6.7 cm
Q

(d)

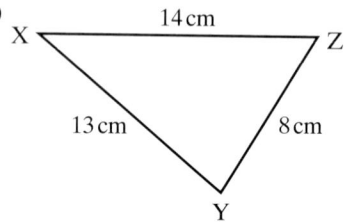

14 cm
X
Z
13 cm
8 cm
Y

3 ABCDEFGH is a cuboid.
ACH is a triangle contained within the cuboid.

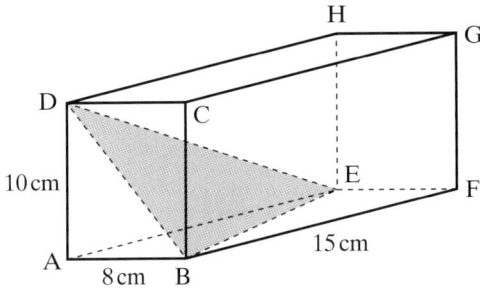

Calculate the size of these angles.
(a) Angle BDE
(b) Angle BED

EXERCISE 40.4H

1 (a) Draw accurately the graph of $y = \sin \theta$ for values of θ from $-180°$ to $180°$.
   (b) For what values of θ in this range does $\sin \theta = 0.7$?

2 (a) Draw accurately the graph of $y = \cos \theta$ for values of θ from $0°$ to $360°$.
   (b) For what values of θ in this range does $\cos \theta = -0.6$?

3 (a) Draw accurately the graph of $y = \sin \theta$ for values of θ from $0°$ to $360°$.
   (b) From your graph read off all the values of θ for which $\sin \theta = -0.8$.

4 (a) Draw accurately the graph of $y = \cos \theta$ for values of θ from $-180°$ to $360°$.
   (b) From your graph read off all the values of θ for which $\cos \theta = -0.3$.

5 One solution to $\sin \theta = 0.6$ is approximately $37°$. Using only the symmetry of the sine curve, find the other angles between $-180°$ and $540°$ that also satisfy the equation $\sin \theta = 0.6$.

6 (a) Draw accurately the graph of $y = \tan \theta$ for values of θ from $-180°$ to $360°$.
   (b) Find, from your graph, the angles for which $\tan \theta = -1.5$.

EXERCISE 40.5H

1 Find the amplitude and the period of each of these curves.
   (a) $y = 2 \sin \theta$ (b) $y = \cos 4\theta$
   (c) $y = 5 \sin 2\theta$ (d) $y = 3 \sin 0.8\theta$
   (e) $y = 4 \cos 5\theta$ (f) $y = 7 \cos 0.6\theta$

2 Draw the graph of $y = 2 \cos \theta$ for values of θ from $0°$ to $360°$.

3 Draw the graph of $y = \sin 3\theta$ for values of θ from $-180°$ to $180°$.

4 Sketch the graph of $y = \cos 5\theta$ for values of θ from $0°$ to $360°$.

5 Sketch the graph of $y = 1.5 \sin \theta$ for values of θ from $-180°$ to $180°$.

6 Find the solutions of $\cos 4\theta = -0.6$ between $0°$ and $360°$.

EXERCISE 41.1H pt

1. $f(x) = x^2 + 4$.
   Find the value of each of these.
   (a) $f(3)$ (b) $f(-2)$

2. $g(x) = x^2 + 2x + 1$.
   Find the value of each of these.
   (a) $g(3)$ (b) $g(-2)$ (c) $g(0)$

3. $h(x) = 5x - 3$
   (a) Solve $h(x) = 7$.
   (b) Write an expression for each of these.
       (i) $h(x - 2)$ (ii) $h(2x)$

4. $f(x) = 2x + 4$
   (a) Solve $f(x) = 1$.
   (b) Write an expression for each of these.
       (i) $2f(x)$ (ii) $f(2x + 3)$

5. $g(x) = 5x - 3$
   (a) Solve $g(x) = 0$.
   (b) Write an expression for each of these.
       (i) $g(x + 3)$ (ii) $2g(x) + 3$

6. $h(x) = x^2 + 2$
   (a) Solve $h(x) = 6$.
   (b) Write down an expression for each of these.
       (i) $h(x + 3)$ (ii) $h(2x) + 1$

7. $f(x) = 2x^2 - 3x$
   (a) Find the value of $f(-2)$.
   (b) Write an expression for each of these.
       (i) $f(x + 2)$ (ii) $f(3x)$

8. $g(x) = x^2 + 3x$
   (a) Solve $g(x) = -2$.
   (b) Write an expression for each of these.
       (i) $3g(x) + 4$ (ii) $g(2x + 1)$

EXERCISE 41.2H

1. (a) Sketch these graphs on the same diagram.
       (i) $y = -x^2$
       (ii) $y = 2 - x^2$
   (b) State the transformation that maps
       $y = -x^2$ on to $y = 2 - x^2$.

2. (a) Sketch these graphs on the same diagram.
       (i) $y = x^2$
       (ii) $y = x^2 - 4$
   (b) State the transformation that maps $y = x^2$
       on to $y = x^2 - 4$.

3. (a) Sketch these graphs on the same diagram.
       (i) $y = x^2$
       (ii) $y = (x - 2)^2$
       (iii) $y = (x - 2)^2 + 3$
   (b) State the transformation that maps $y = x^2$
       on to $y = (x - 2)^2 + 3$.

4. (a) Sketch the result of translating the graph
       of $y = \cos \theta$ by $\begin{pmatrix} 0 \\ 2 \end{pmatrix}$.
   (b) State the equation of the transformed
       graph.

5. State the equation of $y = \tan \theta$ after it has
   been translated by these vectors.
   (a) $\begin{pmatrix} 0 \\ 4 \end{pmatrix}$ (b) $\begin{pmatrix} 3 \\ 0 \end{pmatrix}$

6 The diagram shows the graph of $y = f(x)$.

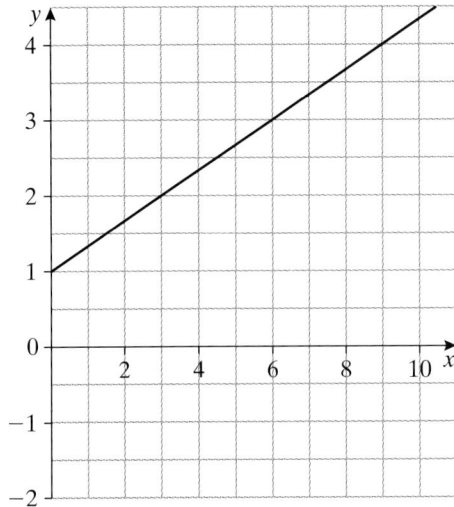

Copy the diagram and sketch these graphs on the same axes.

(a) $y = f(x) - 3$ (b) $y = f(x - 3)$

7 The diagram shows the graph of $y = g(x)$.

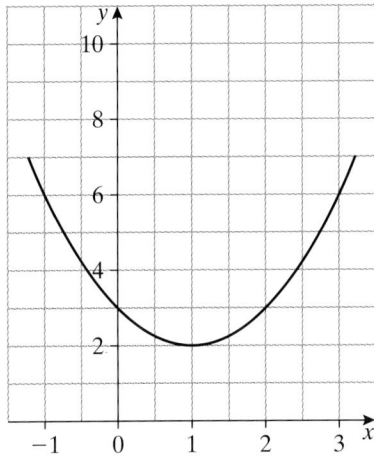

Copy the diagram and sketch these graphs on the same axes.

(a) $y = g(x - 1)$ (b) $y = g(x) - 2$

8 State the equation of the graph $y = x^2$ after it has been translated by $\begin{pmatrix} 3 \\ -4 \end{pmatrix}$.

9 This is the graph of a transformed sine curve. State its equation.

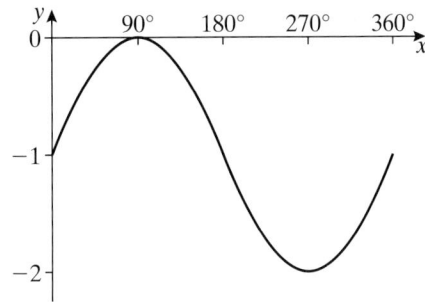

10 The graph of $y = x^2 + 3x$ is translated by $\begin{pmatrix} 2 \\ 3 \end{pmatrix}$.

(a) State the equation of the transformed graph.

(b) Show that this equation can be written as $y = x^2 - x + 1$.

EXERCISE 41.3H

1 (a) Sketch on the same diagram the graphs of $y = \cos \theta$ and $y = 2 \cos \theta$ for $0° \leqslant \theta \leqslant 360°$.

(b) Describe the transformation that maps $y = \cos \theta$ on to $y = 2 \cos \theta$.

2 (a) Sketch on the same diagram the graphs of $y = \sin \theta$ and $y = \sin 2\theta$ for $0° \leqslant \theta \leqslant 360°$.

(b) Describe the transformation that maps $y = \sin \theta$ on to $y = \sin 2\theta$.

3 Describe the transformation that maps

(a) $y = \sin \theta + 1$ on to $y = \sin(-\theta) + 1$.

(b) $y = x^2 + 2$ on to $y = -x^2 - 2$.

(c) $y = x^2$ on to $y = 3x^2$.

(d) $y = \sin \theta$ on to $y = \sin \dfrac{\theta}{3}$.

4 The graph of $y = \sin \theta$ is transformed by a one-way stretch parallel to the θ-axis with a scale factor of $\frac{1}{4}$.

State the equation of the resulting graph.

5 State the equation of the graph $y = x^2 - 1$ after each of these transformations.
   (a) A reflection in the y-axis
   (b) A reflection in the x-axis

6 State the equation of the graph of $y = 4x + 1$ after each of these transformations.
   (a) A one-way stretch parallel to the y-axis with a scale factor of 4
   (b) A one-way stretch parallel to the x-axis with a scale factor of $\frac{1}{2}$

7 Describe the transformation that maps $y = f(x)$ on to each of these graphs.
   (a) $y = f(x) - 2$ **(b)** $y = 3f(x)$
   (c) $y = f(0.5x)$ **(d)** $y = 4f(2x)$

8 $y = x^2 + 3$. Find the equation of the graph after each of these transformations.
   (a) A reflection in the x-axis
   (b) A reflection in the y-axis
   (c) A one-way stretch parallel to the x-axis with a scale factor of 0.5

9 The graph of $y = x^2 - 2x$ is stretched parallel to the x-axis by a scale factor of 2.
   (a) State the equation of the resulting graph.
   (b) What does point $(1, -1)$ map on to under this transformation?

10 The equation of this graph is $y = a \sin b\theta$. Find a and b.

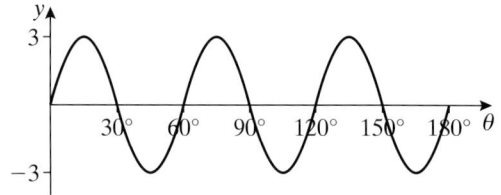

EXERCISE 42.1H

Throughout this exercise, letters in algebraic expressions represent lengths and numbers have no dimensions.

1 Which of these expressions could be a length?
   (a) πd **(b)** $6x^2$ **(c)** $r(\pi + 4)$

2 Which of these expressions could be an area?
   (a) $3ab(4c + d)$ **(b)** $20bh$ **(c)** $a^2 + 6b^2$

3 Which of these expressions could be a volume?
   (a) $2\pi r^3$ **(b)** $4b(c + d)$ **(c)** $\pi r^2 h$

4 State whether each of these expressions represents a length, an area or a volume.
   (a) $2\pi r(r + h)$ **(b)** $\frac{3}{4} \times 2\pi r$ **(c)** $\frac{1}{3}h^3$

5 State whether each of these expressions represents a length, an area, a volume or is nonsense.
   (a) πrh **(b)** $2\pi r(r^2 + 2h)$ **(c)** $\frac{1}{2}(a + b)hp$ **(d)** $\dfrac{12x^3}{ab}$

EXERCISE 42.2H

1 Find the arc length of each of these sectors.
Give your answers to the nearest millimetre.

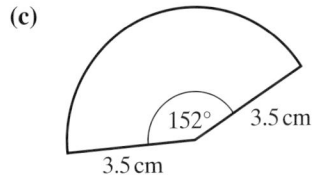

2 Find the area of each of the sectors in question **1**.

3 Find the sector angle of each of these sectors.
Give your answers to the nearest degree.

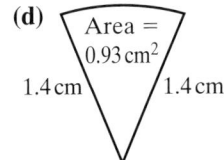

4 Find the radius of each of these sectors.

(a)

5.3 cm

48°

(b)

8.5 cm

167°

(c)

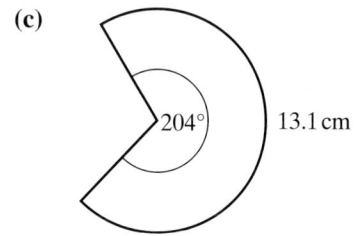

204°

13.1 cm

5 Find the radius of each of these sectors.

(a)

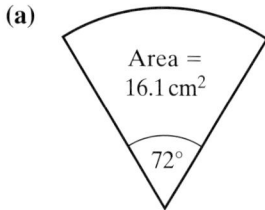

Area = 16.1 cm²

72°

(b)

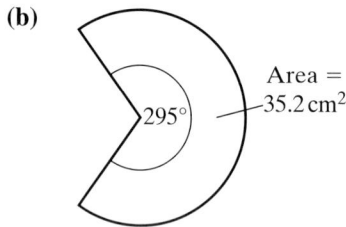

295°

Area = 35.2 cm²

(c)

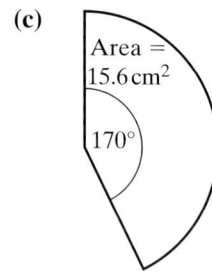

Area = 15.6 cm²

170°

6 A sector of a circle of radius 5.1 cm has area 38 cm².
Calculate the angle of the sector and hence find the arc length of the sector.

7 A decorative frieze is made with alternating
sectors as shown.
The radius of each sector is 75 cm and the
sector angle is 150°.
Calculate the area in square metres covered
by the seven sectors shown.
Give your answer to 3 significant figures.

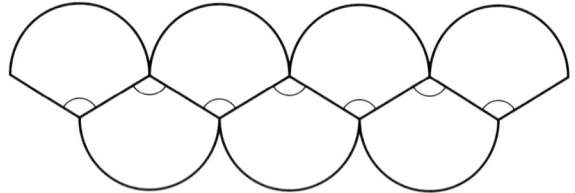

▦ EXERCISE 42.3H (pt)

1 Find the curved surface area of each of these cones.
Give your answers to 3 significant figures.

(a)

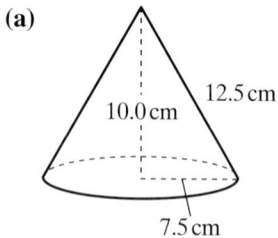

10.0 cm   12.5 cm

7.5 cm

(b)

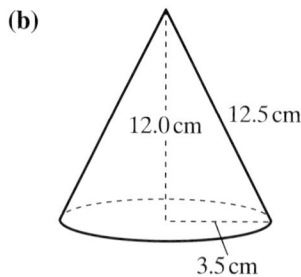

12.0 cm   12.5 cm

3.5 cm

(c)

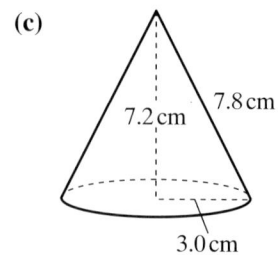

7.2 cm   7.8 cm

3.0 cm

2 Calculate the volume of each of the cones in question **1**.

3 Calculate the volume of each of these square- or rectangular-based pyramids.

(a)

(b)

(c)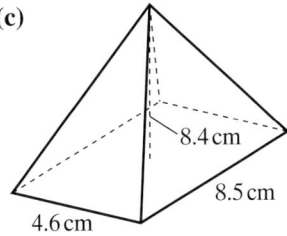

4 A solid cone has a base of radius 6.1 cm and a slant height of 8.4 cm. Calculate its total surface area.

5 A pyramid has a rectangular base of sides 4.5 cm and 6.0 cm. Its volume is 73.8 cm³. Find its height.

6 Find the base radius of each of these cones.
   (a) Volume 256 cm³, height 7.8 cm
   (b) Volume 343 cm³, height 6.5 cm
   (c) Volume 192 cm³, height 10.4 cm

7 A sector of a circle is joined to form a cone.
   Find the radius of the base of the cone made with each of these sectors.
   (a) Radius 8.1 cm, angle 150°
   (b) Radius 10.8 cm, angle 315°
   (c) Radius 7.2 cm, angle 247°

8 Find the surface area of each of these spheres.
   (a) Radius 3.6 cm   (b) Radius 8.5 cm   (c) Diameter 36 cm

9 Find the volumes of each of the spheres in question **8**.

10 Find the radius of each of these spheres.
   (a) Surface area 650 cm²   (b) Surface area 270 cm²
   (c) Volume 1820 cm³   (d) Volume 3840 cm³

EXERCISE 42.4H

1 Calculate the area of each of the shaded segments.

(a)

(b)

(c)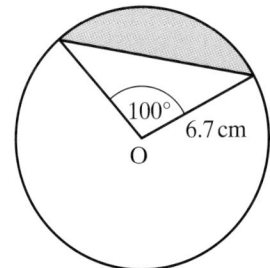

2 A concrete gate-post is in the shape of a cuboid 1.6 m by 30 cm by 30 cm topped with a sphere of radius 12 cm.
Calculate the volume of the gate-post.

3 Calculate the area of each of the shaded major segments.

(a)

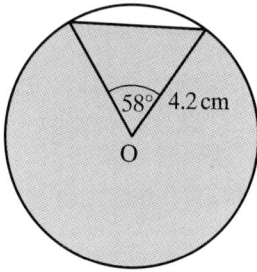

58° 4.2 cm
O

(b)

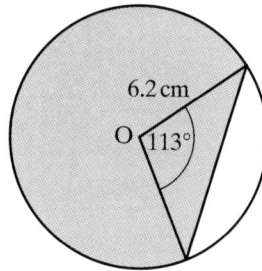

6.2 cm
O 113°

(c)

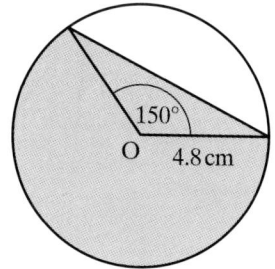

150°
O 4.8 cm

4 A birthday cake has radius 9 cm and height 8 cm. Its top and sides are covered in icing.
Marie is given a slice of the cake. It is a sector of angle 50°.
Calculate the surface area of the icing on Marie's slice of cake.

5 Calculate the perpendicular height of each of these cones and hence find their volumes.

(a)

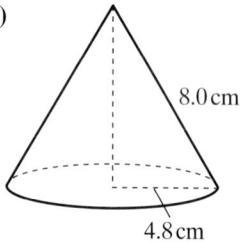

8.0 cm
4.8 cm

(b)

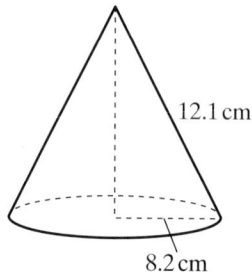

12.1 cm
8.2 cm

(c)

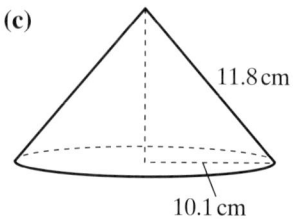

11.8 cm
10.1 cm

6 A solid cone has a base of radius 7.9 cm and a height of 11.8 cm.
(a) Calculate its volume.
(b) Find its slant height and hence its total surface area.

7 The top half in height of a cone is removed.
Show that the volume of the remaining frustum is $\frac{7}{8}$ of the volume of the original cone.
What fraction of the original curved surface area of the cone is the curved surface area of the frustum?

8 A podium is a frustum of a cone. The height of the frustum is 1.2 m and the diameters of its top and its base are 0.8 m and 1.4 m respectively.
(a) Show that the podium is a frustum of a cone of complete height 2.8 m.
(b) Calculate the volume of the podium.

9 A pyramid has a square base of side 12.8 cm and its volume is 524 cm³.
(a) Calculate its height.
(b) Hence show that the sloping edges (which are all of equal length) are 13.2 cm long.

10 A lampshade is made from a piece of parchment.
(a) Find in the form $\frac{k}{\pi}$ the angle of the sector of the inner circle which has been removed.
(b) Hence calculate the surface area of the lampshade.

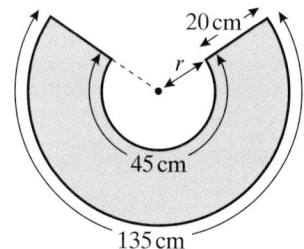

20 cm
r
45 cm
135 cm

EXERCISE 43.1H 🔵

1   A bag contains five red counters, three green counters and two yellow counters.
    What is the probability of selecting
    (a) a red ball or a green ball?
    (b) a green ball or a yellow ball?

2   When Mr Smith goes on holiday the probability that he goes to the seaside is 0.4, to the countryside is 0.35 and to a city is 0.25.
    What is the probability that he goes on holiday
    (a) to the countryside or the city?
    (b) to the seaside or the city?

3   A spinner is numbered 1 to 5.
    The probabilities of each of the numbers occurring are given in the table.

Number	1	2	3	4	5
Probability	0.39	0.14	0.22	0.11	0.14

    What is the probability that in one spin the number will be
    (a) 1 or 2?   **(b)** 3, 4 or 5?
    (c) an odd number?   **(d)** less than 2?
    (e) at least 3?

4   There are four kings and four aces in a pack of 52 playing cards.
    A card is chosen at random.
    What is the probability that it is a king or an ace?

5   A coin is tossed and a dice is thrown.
    What is the probability of getting a head on the coin and an even number on the dice?

6   An ordinary dice is thrown three times.
    What is the probability that the dice lands on 6 each time?

7   The probability that the school team wins their next hockey match is 0.8.
    What is the probability that, in their next two matches, the school team
    (a) wins both matches?
    (b) wins neither match?

8   Each of the letters of the word MISSISSIPPI is written on a card.
    The cards are shuffled and one is selected.
    This card is returned to the pack which is again shuffled.
    A second card is selected.
    What is the probability that the two cards are
    (a) both P?
    (b) both S?
    (c) both a consonant?

9   A box contains a large number of red beads and a large number of white beads.
    40% of the beads are red.
    A bead is chosen from the box, its colour is noted and it is replaced.
    A second bead is then chosen.
    What is the probability that
    (a) both beads are red?
    (b) both beads are white?
    (c) one bead of each colour is chosen?

10 In a large batch of light bulbs, the probability that a bulb is defective is 0.01.
    Three light bulbs are selected at random for testing.
    What is the probability that
    (a) all three work?
    (b) all three are defective?
    (c) two of the three are defective?

EXERCISE 43.2H 🅿️

1 Pat is playing a board game with a spinner numbered 1 to 4.
She spins the spinner twice.
   (a) Copy and complete the tree diagram.

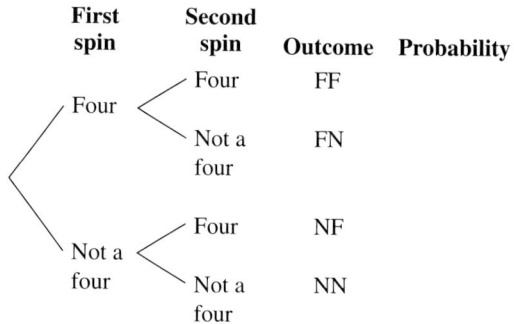

First spin	Second spin	Outcome	Probability
Four	Four	FF	
	Not a four	FN	
Not a four	Four	NF	
	Not a four	NN	

   (b) Use the tree diagram to work out the probability that
      (i) Pat gets two 4s.   **(ii)** Pat gets just one 4.

2 The probability that I get up late on any day is 0.3.
   (a) Draw a tree diagram to show my getting up late or not late on two days.
   (b) Work out the probability that
      (i) I don't get up late on either of the two days.
      (ii) I get up late on one of the two days.

3 There are five red discs and three blue discs in a bag.
A disc is selected, its colour is noted and it is then replaced in the bag.
A second disc is then selected.
   (a) Draw a probability tree diagram to show the outcomes of the two selections.
   (b) Use the tree diagram to find the probability that
      (i) both discs are red.
      (ii) both discs are the same colour.
      (iii) at least one disc is blue.

4 The probabilities that Phil wins, draws or loses any game of chess are 0.6, 0.3 and 0.1 respectively.
   (a) Draw a tree diagram to show the outcomes of Phil's next two games.
   (b) Work out the probability that
      (i) Phil wins both games.
      (ii) Phil wins one of the two games.
      (iii) the results of the two games are the same.

5 On the way to work, Beverley passes through three sets of traffic lights.
The probability that the first set of lights is green when she reaches them is 0.6.
The probability that the second set is green is 0.7.
The probability that the third set is green is 0.8.
   (a) Draw a probability tree diagram to show the possible outcomes.
   (b) What is the probability that she has to stop at
      (i) all three sets of lights?
      (ii) just one set of lights?
      (iii) at least two sets of lights?

EXERCISE 43.3H (pt)

1   There are four black beads and three white beads in a box.
    A bead is selected at random and not replaced.
    A second bead is then selected.
    (a) Draw a probability tree diagram to show all the possible outcomes.
    (b) Find the probability that the two beads are
        (i) both white.
        (ii) one of each colour.

2   Ben randomly selects three cards, without replacing any, from a normal pack of 52 playing cards.
    What is the probability that
    (a) all three are aces?
    (b) two of the three are aces?

3   The probability of it snowing one day in winter is reported to be 0.2.
    If it snows on that day, the probability that it snows the following day is 0.7.
    If it doesn't snow, the probability that it will snow the following day is 0.1.
    (a) Draw a probability tree diagram to show the possible outcomes.
    (b) What is the probability of
        (i) it snowing on both days?
        (ii) it not snowing on both days?
        (iii) it snowing on at least one of the two days?

4   When Elaine goes to school, she either walks, cycles or goes by bus.
    The probability that she walks is 0.5 and that she cycles is 0.2.
    If she walks the probability that she is late is 0.4, if she cycles it is 0.1, and if she goes by bus it is 0.2.
    What is the probability that she is late for school?

5   There are nine boys and fifteen girls in a class.
    Three children are to be randomly selected to represent the class in a competition.
    What is the probability that
    (a) all three are girls?
    (b) one is a girl and two are boys?
    (c) at least two girls are chosen?

44 → VECTORS

EXERCISE 44.1H

1 On squared paper, draw and label vectors to represent these column vectors.
Do not forget to put the arrows on.

$$\mathbf{a} = \begin{pmatrix} 3 \\ 6 \end{pmatrix}, \qquad \mathbf{b} = \begin{pmatrix} 5 \\ -2 \end{pmatrix}, \qquad \mathbf{c} = \begin{pmatrix} -4 \\ 1 \end{pmatrix},$$

$$\mathbf{d} = \begin{pmatrix} 0 \\ 3 \end{pmatrix}, \qquad \mathbf{e} = \begin{pmatrix} -4 \\ 0 \end{pmatrix}, \qquad \mathbf{f} = \begin{pmatrix} -5 \\ -3 \end{pmatrix}$$

2 Write down column vectors to represent these vectors.

(a) $\overrightarrow{AB}$ (b) $\overrightarrow{CB}$ (c) $\overrightarrow{BC}$

(d) $\overrightarrow{DC}$ (e) $\overrightarrow{DE}$ (f) $\overrightarrow{ED}$

(g) $\overrightarrow{AD}$ (h) $\overrightarrow{EA}$

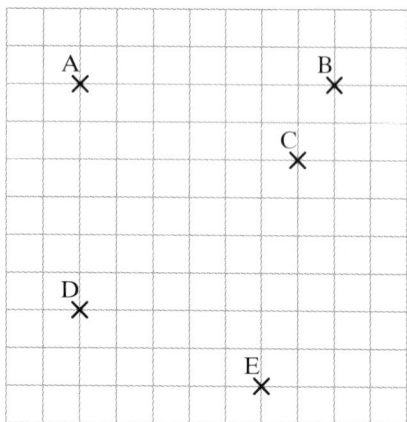

3 Find the column vector that maps the point
(a) (4, 3) on to (4, 7).
(b) (1, 3) on to (6, 7).
(c) (1, 7) on to (5, 1).
(d) (5, −2) on to (3, 6).
(e) (4, 7) on to (−1, 3).
(f) (7, −8) on to (−3, −3).

4 Write each of these vectors in terms of **a** and/or **b** as shown in the diagram.

(a) $\overrightarrow{AB}$ (b) $\overrightarrow{CD}$ (c) $\overrightarrow{EF}$

(d) $\overrightarrow{GH}$ (e) $\overrightarrow{IJ}$ (f) $\overrightarrow{KL}$

(g) $\overrightarrow{MN}$ (h) $\overrightarrow{PQ}$

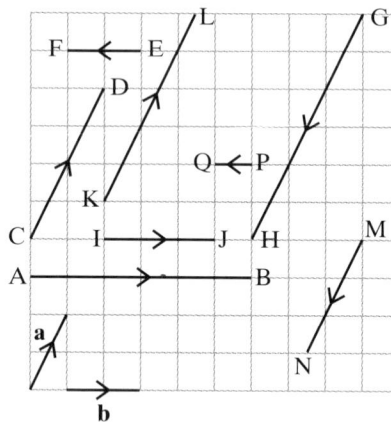

5 ABCDEF is a regular hexagon.
O is the centre of the hexagon.
$\overrightarrow{OA} = \mathbf{a}$ and $\overrightarrow{OB} = \mathbf{b}$

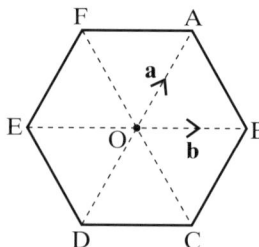

Write down each of these vectors in terms of **a** and/or **b**.

(a) $\overrightarrow{DC}$ (b) $\overrightarrow{DA}$ (c) $\overrightarrow{AF}$

(d) $\overrightarrow{BE}$ (e) $\overrightarrow{BC}$ (f) $\overrightarrow{AD}$

6 Copy and complete this table.

	Original point	Vector	New point
(a)	(1, 3)	$\begin{pmatrix} 2 \\ 5 \end{pmatrix}$	
(b)	(6, 1)	$\begin{pmatrix} -2 \\ 3 \end{pmatrix}$	
(c)	(8, 3)	$\begin{pmatrix} -5 \\ -3 \end{pmatrix}$	
(d)	(7, −2)	$\begin{pmatrix} -4 \\ 5 \end{pmatrix}$	
(e)	(−6, −1)	$\begin{pmatrix} -2 \\ 5 \end{pmatrix}$	
(f)	(−3, 6)	$\begin{pmatrix} -1 \\ -8 \end{pmatrix}$	

EXERCISE 44.2H

1 The diagram shows the vectors **a** and **b**.

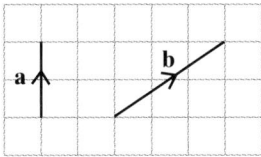

(a) Draw vectors **a** and **b** on squared paper.
(b) Draw vectors to represent each of these.
    (i) **a** + **b**
    (ii) **b** − **a**
    (iii) 2**a** + **b**
    (iv) 2**a** − **b**
    (v) **a** + $\frac{1}{2}$**b**

2 $\mathbf{a} = \begin{pmatrix} 3 \\ 5 \end{pmatrix}$, $\mathbf{b} = \begin{pmatrix} -1 \\ 4 \end{pmatrix}$ and $\mathbf{c} = \begin{pmatrix} 0 \\ 2 \end{pmatrix}$.

Work out these.
(a) **a** + **b**   (b) **a** − **b**
(c) 2**a** + 3**b**   (d) 2**a** − **c**
(e) **b** + 2**a** − $\frac{1}{2}$**c**   (f) 2**a** − 3**b** + 5**c**

3 ABCD is a parallelogram.
E, F, G and H are the midpoints of the sides.
$\overrightarrow{AB} = \mathbf{p}$ and $\overrightarrow{AD} = \mathbf{q}$

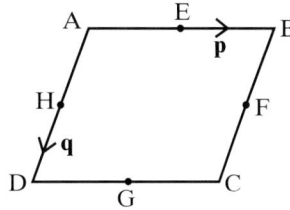

Find each of these vectors in terms of **p** and **q**.
(a) $\overrightarrow{AC}$ (b) $\overrightarrow{AF}$ (c) $\overrightarrow{ED}$
(d) $\overrightarrow{CE}$ (e) $\overrightarrow{FG}$

4 ABCD is a trapezium.
$\overrightarrow{AD} = \mathbf{p}$ and $\overrightarrow{AB} = \mathbf{q}$
$\overrightarrow{BC} = \frac{1}{2}\overrightarrow{AD}$

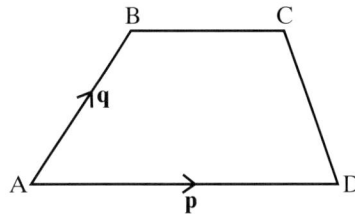

Find each of these vectors, as simply as possible, in terms of **p** and/or **q**.
(a) $\overrightarrow{BC}$ (b) $\overrightarrow{AC}$
(c) $\overrightarrow{BD}$ (d) $\overrightarrow{DC}$

5 In the diagram, P is two thirds of the way along AB.
$\overrightarrow{OA} = \mathbf{a}$ and $\overrightarrow{OB} = \mathbf{b}$

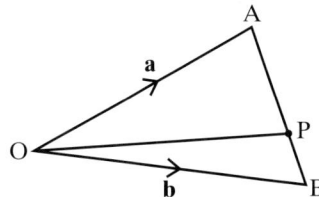

Find each of these vectors, as simply as possible, in terms of **a** and **b**.
(a) $\overrightarrow{BA}$ (b) $\overrightarrow{BP}$ (c) $\overrightarrow{OP}$

6 In the diagram, AOD and BOC are straight lines.

OD is 3 times as long as AO.
OC is 3 times as long as BO.
$\overrightarrow{OA} = \mathbf{a}$ and $\overrightarrow{OB} = \mathbf{b}$

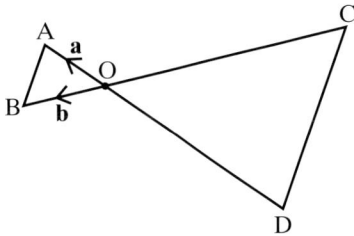

(a) Find each of these vectors in terms of **a** and **b**.

(i) $\overrightarrow{AB}$ **(ii)** $\overrightarrow{OD}$

(iii) $\overrightarrow{OC}$ **(iv)** $\overrightarrow{CD}$

(b) What two facts can you conclude about the lines AB and CD?

7 Write each of these vectors in terms of **a** and **b**.

(a) $\overrightarrow{AB}$ **(b)** $\overrightarrow{CD}$ **(c)** $\overrightarrow{EF}$

(d) $\overrightarrow{GH}$ **(e)** $\overrightarrow{IJ}$ **(f)** $\overrightarrow{KL}$

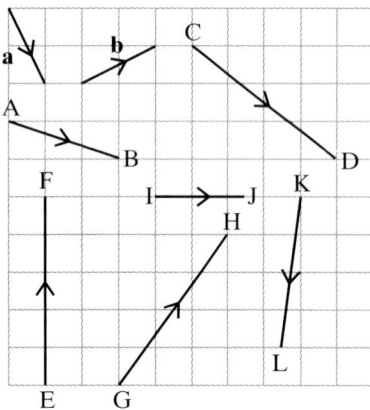

8 In the diagram, $\overrightarrow{OA} = \mathbf{a}$ and $\overrightarrow{OC} = \mathbf{c}$.
D is one third of the way along AC.

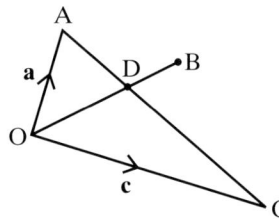

(a) Find each of these vectors, as simply as possible, in terms of **a** and **c**.

(i) $\overrightarrow{AC}$ **(ii)** $\overrightarrow{AD}$ **(iii)** $\overrightarrow{OD}$

(b) $\overrightarrow{OB} = \frac{3}{2} \overrightarrow{OD}$

Find each of these vectors, as simply as possible, in terms of **a** and **c**.

(i) $\overrightarrow{OB}$ **(ii)** $\overrightarrow{AB}$

(c) What two facts can you conclude about the lines AB and OC?

EXERCISE 45.1H

Simplify these.

1 $\dfrac{x+4}{3} + \dfrac{x-1}{2}$

2 $\dfrac{x+5}{3} - \dfrac{x+3}{4}$

3 $\dfrac{x-3}{2} - \dfrac{x-5}{3}$

4 $\dfrac{4x+3}{2} - \dfrac{3x-2}{4}$

5 $\dfrac{2}{x+3} + \dfrac{x-1}{3}$

6 $\dfrac{2}{x-5} + \dfrac{x+4}{2}$

7 $\dfrac{3x+4}{5} - \dfrac{3}{2x-1}$

8 $\dfrac{5x+4}{3x-2} + \dfrac{5}{4}$

EXERCISE 45.2H 🔵pt

Simplify these.

1 $\dfrac{3}{x+2} + \dfrac{2}{x-1}$

2 $\dfrac{4}{x-3} + \dfrac{x+3}{2x}$

3 $\dfrac{x+5}{2x} - \dfrac{3x}{x+2}$

4 $\dfrac{x+5}{x-3} - \dfrac{x-3}{x+2}$

5 $\dfrac{1}{x+4} - \dfrac{3}{x-5}$

6 $\dfrac{x}{x-3} + \dfrac{x-2}{x+1}$

7 $\dfrac{3x+2}{x-5} - \dfrac{x-3}{x-4}$

8 $\dfrac{2x+3}{3x+1} - \dfrac{x-2}{2x+5}$

EXERCISE 45.3H

Solve these.

1 $\dfrac{x-3}{2} - \dfrac{x-2}{3} = 1$

2 $\dfrac{5}{x-4} - \dfrac{3}{x-2}$

3 $\dfrac{1}{4x-3} = \dfrac{1}{3x+2}$

4 $\dfrac{3-x}{4} + \dfrac{2x+5}{3} = 1$

5 $\dfrac{x-2}{5} - \dfrac{2x-3}{4} = \dfrac{1}{3}$

6 $\dfrac{3x}{x+4} + \dfrac{2x}{5x-2} = \dfrac{3}{2}$

7 $\dfrac{2}{3x+1} - \dfrac{5}{x+3} = 0$

8 $\dfrac{3}{x-2} - \dfrac{1}{x+1} = 1$

EXERCISE 46.1H (pt)

In each of the questions, find the size of the angle or length marked with a letter.
Give a reason for each step of your work.

1

2

3

4

5

6

7

8

9

10

11

12

13

14

15

16

17

18

19

20

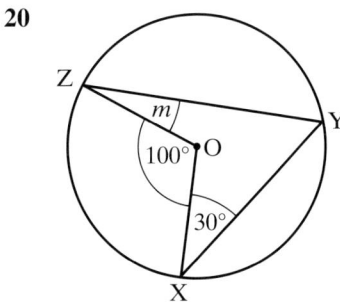

EXERCISE 46.2H pt

In each of the questions, find the size of the angles marked with a letter.
Give a reason for each step of your work.

1

2

3

4

5

6

7

8

9

10

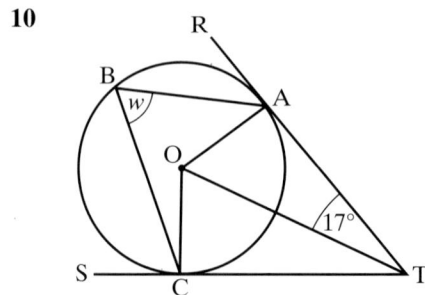